Federal Aviation Administration
Office of Environment and Energy

FAA-EE-2005-01

SAGE
System for assessing Aviation's Global Emissions

Version 1.5

Technical Manual

Brian Kim, Gregg Fleming, Sathya Balasubramanian, Andrew Malwitz, Joosung Lee, and Joseph Ruggiero
Volpe National Transportation Systems Center
Environmental Measurements and Modeling Division
Cambridge, MA

Ian Waitz, Kelly Klima
Massachusetts Institute of Technology
Department of Aeronautics and Astronautics
Cambridge, MA

Virginia Stouffer, Dou Long, and Peter Kostiuk
Logistics Management Institute
McLean, Virginia

Maryalice Locke, Curtis Holsclaw, Angel Morales, Edward McQueen, and Warren Gillette
Federal Aviation Administration
Office of Environment and Energy
Washington, DC

September 2005

NOTICE

This document is disseminated under the sponsorship of the Department of Transportation in the interest of information exchange. The United States Government assumes no liability for its contents or use thereof. This report does not constitute a standard, specification, or regulation.

The United States Government does not endorse products or manufacturers. Trade or manufacturers' names appear herein solely because they are considered essential to the object of this document.

REPORT DOCUMENTATION PAGE		Form Approved OMB No. 0704-0188

Public reporting burden for this collection of information is estimated to average one hour per response, including the time for reviewing instructions, searching existing data sources, gathering and maintaining the data needed, and completing and reviewing the collection of information. Send comments regarding this burden estimate or any other aspect of this collection of information, including suggestions for reducing this burden, to Washington Headquarters Services, Directorate for Information Operations and Reports, 1215 Jefferson Davis Highway, Suite 1204, Arlington, VA 22202-4302, and to the Office of Management and Budget, Paperwork Reduction Project (0704-0188), Washington, DC 20503.

1. AGENCY USE ONLY (Leave blank)	2. REPORT DATE September 2005	3. REPORT TYPE AND DATES COVERED Final Report September 2005
4. TITLE AND SUBTITLE System for assessing Aviation's Global Emissions (SAGE), Version 1.5, Technical Manual		5. FUNDING NUMBERS FA5N/BS043
6. AUTHOR(S) Brian Kim[1], Gregg Fleming[1], Sathya Balasubramanian[1], Andrew Malwitz[1], Joosung Lee[1], Joseph Ruggiero[1], Ian Waitz[2], Kelly Klima[2], Virginia Stouffer[3], Dou Long[3], Peter Kostiuk[3], Maryalice Locke[4], Curtis Holsclaw[4], Angel Morales[4], Edward McQueen[4], Warren Gillette[4]		
7. PERFORMING ORGANIZATION NAME(S) AND ADDRESS(ES) (1) U.S. Department of Transportation Research and Innovative Technology Administration John A. Volpe National Transportation Systems Center Environmental Measurement and Modeling Division DTS-34 55 Broadway Cambridge, MA 02142 (2) Massachusetts Institute of Technology Department of Aeronautics and Astronautics Cambridge, MA 02142 (3) Logistics Management Institute 2000 Corporate Ridge McLean, Virginia 22102		8. PERFORMING ORGANIZATION REPORT NUMBER DOT-VNTSC-FAA-05-14
9. SPONSORING/MONITORING AGENCY NAME(S) AND ADDRESS(ES) (4) Federal Aviation Administration Office of Environment and Energy 800 Independence Ave., S.W. Washington, DC 20591		10. SPONSORING/MONITORING AGENCY REPORT NUMBER FAA-EE-2005-01
11. SUPPLEMENTARY NOTES: FAA AEE Emissions Division Office Manager: Curtis Holsclaw FAA AEE Program Manager: Maryalice Locke		
12a. DISTRIBUTION/AVAILABILITY STATEMENT: Publicly Available		12b. DISTRIBUTION CODE

13. ABSTRACT (Maximum 200 words)

The United States (US) Federal Aviation Administration (FAA) Office of Environment and Energy (AEE) has developed the System for assessing Aviation's Global Emissions (SAGE) with support from the Volpe National Transportation Systems Center (Volpe), the Massachusetts Institute of Technology (MIT) and the Logistics Management Institute (LMI). Currently at Version 1.5, SAGE is a high fidelity computer model used to predict aircraft fuel burn and emissions for all commercial (civil) flights globally in a given year. This means that the model is capable of analyzing scenarios at from a single flight to airport, country, regional, and global levels. SAGE is able to dynamically model aircraft performance, fuel burn and emissions, capacity and delay at airports, and forecasts of future scenarios. FAA's purpose in developing SAGE is to provide the international aviation community with a tool to evaluate the effects of various policy, technology, and operational scenarios on aircraft fuel use and emissions. FAA is committed to the continued development and support of SAGE. Although the results from the model have been made available to the international aviation community, SAGE is currently an FAA government research tool and has not been released to the general public.

As part of SAGE development, this document has been prepared to provide detailed specifications of the model's methods, data, and assumptions. The information is provided within the context of using SAGE to develop global inventories of fuel burn and emissions.

14. SUBJECT TERMS SAGE, aircraft, fuel burn, emissions, emissions inventory, aircraft performance, global flights, computer model, emissions model, world fleet, forecasting, capacity and delay, aircraft movements			15. NUMBER OF PAGES 234
			16. PRICE CODE
17. SECURITY CLASSIFICATION OF REPORT Unclassified	18. SECURITY CLASSIFICATION OF THIS PAGE Unclassified	19. SECURITY CLASSIFICATION OF ABSTRACT Unclassified	20. LIMITATION OF ABSTRACT

TABLE OF CONTENTS

1 INTRODUCTION 1
   1.1 Background..1
   1.2 Objective and Scope 2
   1.3 Modeling Capabilities..2
   1.4 Document Outline..2
2 DATABASE DESCRIPTIONS 4
   2.1 ICAO Emissions Databank...4
   2.2 Emissions and Dispersion Modeling System (EDMS) Emissions Data......................................5
   2.3 Base of Aircraft Data (BADA)...5
   2.4 Integrated Noise Model (INM) and the Society of Automotive Engineers (SAE) Aerospace
       Information Report (AIR) 1845...6
   2.5 Airports..7
   2.6 Official Airline Guide (OAG) 8
   2.7 Enhanced Traffic Management System (ETMS)..9
   2.8 BACK World Fleet Registration Database 11
   2.9 ICAO's Forecasting and Economics Sub-Group's (FESG) Forecasts 13
   2.10 FAA's Terminal Area Forecasts (TAF)..14
   2.11 Airline On-Time Performance Data 14
   2.12 Airport Capacity 15
   2.13 Aircraft Phase-Out Indicators and Fractions 15
   2.14 Aircraft Retirement Parameters 16
3 METHODOLOGY – HISTORICAL INVENTORY...17
   3.1 Process Flight Data...20
      3.1.1 ETMS Processing..20
         3.1.1.1 Extraction..20
         3.1.1.2 Compression 20
         3.1.1.3 Filtering..22
         3.1.1.4 Flight Plan Cleaning 22
         3.1.1.5 Trajectory Cleaning 23
         3.1.1.6 Preliminary Computations..24
      3.1.2 OAG Processing..24
         3.1.2.1 Expansion..24
         3.1.2.2 Filtering..24
   3.2 Create Aircraft Movements Database...25
      3.2.1 Development of Aircraft and Engine Mapping Tables 25
      3.2.2 Conduct Engine Assignments...26
      3.2.3 Develop Cruise Altitudes and Trajectories..27
      3.2.4 Develop Taxi Times..28
      3.2.5 Calculate Delays...29
      3.2.6 Finalize Movements Database..29
   3.3 Preparation of Data for Fuel Burn and Emissions Calculations 31
      3.3.1 Trip Distance/Takeoff Weights...31
      3.3.2 Aircraft Performance Data 32
      3.3.3 Emissions Data...33
      3.3.4 Unscheduled and Canceled Flights Scaling Factors...33
   3.4 Calculate Fuel Burn and Emissions..35
      3.4.1 Atmospheric data..35

3.4.2 Aircraft Performance Modeling 36
   3.4.2.1 Calculate Takeoff/Climbout and Approach Trajectories 36
   3.4.2.2 Determine Aircraft Speed Schedule 37
   3.4.2.3 Calculate Lift and Drag 38
   3.4.2.4 Calculate Thrust and Rate of Climb 39
3.4.3 Fuel Flow and Fuel Burn Computations 41
3.4.4 Emissions Modeling 42
   3.4.4.1 Modeling CO, HC, and NOx Emissions 42
   3.4.4.2 Modeling CO_2, H_2O, and SOx Emissions 46
3.4.5 Additional Data Checks 46

4 METHODOLOGY – FORECASTING 47
4.1 Growth of Scheduled Flights 48
4.1.1 TAF to Grow US Flights 48
4.1.2 FESG to Grow Non-US Flights 49
4.1.3 Modeling in WWLMINET 49
4.2 Aircraft Retirements and Replacements 49
4.2.1 Create Baseline Fleet 49
4.2.2 Create Future Aircraft 50
4.2.3 Creation of Forecasted Fleet 50
4.2.4 Application of Retirement Parameters 50
4.2.5 Development of Conversion Factors 51
4.2.6 Application of Conversion Factors 51
4.3 Application of Baseline and Forecasted Schedules 52
4.3.1 Comparative analyses 52
4.3.2 Development of Future Magnitudes 52

5 OUTPUT RESULTS 53
5.1 SAGE Raw Inventories 53
5.1.1 Raw Flight-Level Modal Inventory 53
5.1.2 Raw Chord-Level Inventory 55
5.1.3 Raw 4D World Gridded Inventory 56
5.2 Processed Inventories 57

6 CONCLUSIONS 63
REFERENCES 64
APPENDIX A: Country-Region Mapping Tables 68
APPENDIX B: Analysis of Duplicate Flights in OAG 71
APPENDIX C: Helicopters Currently Removed from the SAGE Movements Database 72
APPENDIX D: Aircraft Code Mappings used for Aircraft Performance Modeling 73
APPENDIX E: Aircraft Code Mappings used for Engine Assignments 84
APPENDIX F: Engine Mappings 113
APPENDIX G: Cruise Altitude Dispersion Distributions 121
APPENDIX H: Sample Cruise Track Dispersion Distributions 124
APPENDIX I: WWLMINET 131
APPENDIX J: Uncertainty of Scaled SAGE Inventories using Sampled Datasets 142
APPENDIX K: Aircraft Performance Data 147
APPENDIX L: A Methodology to Account for Unscheduled and Canceled Commercial Flights 149
APPENDIX M: Fuel Burn and Emissions Module 191
APPENDIX N: ICAO Engines Affected by Anomalous ICAO Data 219
APPENDIX O: Forecasting Pseudo-Code 222

LIST OF ACRONYMS AND ABBREVIATIONS

4D	Four-Dimensional
ACI	Airport Council International
AEE	Office of Environment and Energy
AIR	Aerospace Information Report
ANCAT	Abatement of Nuisances Caused by Air Transport
ANCAT/EC2	Expert European group on global emissions inventory development
APO	Office of Aviation Policy and Plans
ARTCC	Air Route Traffic Control Center
ARTS	Automatic Radar Tracking System
ASQP	Airline Service Quality Performance
ATC	Air Traffic Control
ATCSCC	Air Traffic Control System Command Center
ATM	Air Traffic Management
BACK	BACK Aviation Solutions
BADA	Base of Aircraft Data
BFFM2	Boeing Fuel Flow Method 2
BTS	Bureau of Transportation Statistics
CAEP	Committee on Aviation Environmental Protection
CAS	Calibrated Air Speed
CDA	Continuous Descent Approach
CNS	Communication, Navigation, and Surveillance
CRS	Computer Reservation System
DLR	Deutsche Forschungsanstalt fur Luft- and Raumfahrt
EC	European Comission
ECAC	European Civil Aviation Conference
EDMS	Emissions and Dispersion Modeling System
EI	Emissions Index
ESRI	Environmental Systems Research Institute
ETMS	Enhanced Traffic Management System
Eurocontrol	European Organization for the Safety of Air Navigation
FAA	Federal Aviation Administration
FBE	Fuel Burn and Emissions (module)
FESG	Forecasting and Economics Sub Group
GC	Great Circle
GDP	Gross Domestic Product
GIS	Geographic Information System
GMT	Greenwich Mean Time
GSE	Ground Service Equipment
IATA	International Air Transport Association
ICAO	International Civil Aviation Organization
ID	Identification (e.g., ID number)
ILS	Instrument Landing System
INM	Integrated Noise Model
IPCC	Intergovernmental Panel on Climate Change
ISA	International Standard Atmosphere
LMI	Logistics Management Institute
LMINET	Logistics Management Institute network queuing model of the US

LTO	Landing and Takeoff
MAGENTA	Model for Assessment of Global Exposure due to Noise from Transport Aircraft
MIT	Massachusetts Institute of Technology
MSL	Mean Sea Level
MTF	Mixed Turbofan
MMU	Manchester Metropolitan University
NASA	National Aeronautics and Space Administration
NPD	Noise versus Power versus Distance
NPIAS	National Plan of Integrated Airport System
OAG	Official Airline Guide
OD	Origin-Destination
PM	Particulate Matter
ROC	Rate of Climb
RVSM	Reduced Vertical Separation Minimum
SAGE	System for assessing Aviation's Global Emissions
SAE	Society of Automotive Engineers
SI	System International
TAF	Terminal Area Forecast
TAS	True Air Speed
TEM	Total Energy Model
TERP	Terminal and En Route Procedure
TF	Turbofan
TIM	Time In Mode
TMU	Traffic Management Unit
TOGW	Takeoff Gross Weight
TRACON	Terminal Radar Approach Control
TSFC	Thrust Specific Fuel Consumption
OD	Origin-Destination
UID	Unique Identification Number
UN	United Nations
UNFCCC	United Nations Framework Convention on Climate Change
US	United States
USDOT	United States Department of Transportation
USEPA	United States Environmental Protection Agency
WG3	Working Group 3
WWLMINET	Worldwide version of the LMINET

LIST OF FIGURES

Figure 1. Full Gate-to-Gate Flight Modeling in SAGE 17
Figure 2. Example Takeoff and Climbout Flight Chords 17
Figure 3. Overview of the Main Modules and Databases in SAGE 19
Figure 4. Trajectory Compression 21
Figure 5. Cruise Altitude and Track Dispersions 28
Figure 6. Takeoff/Climbout and Approach Trajectories 36
Figure 7. Log-Log Plots of EI versus Fuel Flow for NOx, CO, and HC 43
Figure 8. Overview of Forecasting within SAGE 47
Figure 9. Orientation of Standard Grids within SAGE 57
Figure 10. Example Regional Inventory 58
Figure 11. Example Country Inventory for Selected Countries 59
Figure 12. Example Aircraft-Engine Inventory for Selected Equipment 60
Figure 13. Example Fuel Burn Plot of 1° Latitude by 1° Longitude World Grids with all Altitudes aggregated 61
Figure 14. Worldwide Airport Locations Color-Coded by Region 62

LIST OF TABLES

Table 1. Typical 10-Year Projection of Number of Aircraft as a Function of Generic Seat Class 13
Table 2. Aircraft Survivor Curve Coefficients 16
Table 3. Example Engine Distribution 26
Table 4. Example Table of Aircraft Takeoff Weights as a Function of Trip Distance 31
Table 5. Modal Drag Coefficient Scaling Factors 32
Table 6. Scaling Factors for Outlier Airports 34
Table 7. Issues Associated with the use of BFFM2 and the ICAO Emissions Databank 45

1 INTRODUCTION

The United States (US) Federal Aviation Administration (FAA) Office of Environment and Energy (AEE) has developed the System for assessing Aviation's Global Emissions (SAGE) with support from the Volpe National Transportation Systems Center (Volpe), the Massachusetts Institute of Technology (MIT) and the Logistics Management Institute (LMI). Currently at Version 1.5, SAGE is a high fidelity computer model used to predict aircraft fuel burn and emissions for all commercial (civil) flights globally in a given year. This means that the model is capable of analyzing scenarios at from a single flight to airport, country, regional, and global levels. SAGE is able to dynamically model aircraft performance, fuel burn and emissions, capacity and delay at airports, and forecasts of future scenarios. FAA's purpose in developing SAGE is to provide the international aviation community with a tool to evaluate the effects of various policy, technology, and operational scenarios on aircraft fuel use and emissions. FAA is committed to the continued development and support of SAGE. Although the results from the model have been made available to the international aviation community, SAGE is currently an FAA government research tool and has not been released to the general public.

As part of SAGE development, this document has been prepared to provide detailed specifications of the model's methods, data, and assumptions. The information is provided within the context of using SAGE to develop global inventories of fuel burn and emissions. Since the purpose of this document is to present the technical details of the model, the outputs and validation assessment discussions are generally not included. Such discussions can be found in FAA[c, d] 2005.

1.1 Background

The development of SAGE was in part stimulated by the rapid growth in aviation and the need for better emissions modeling capabilities on a global level. According to the "Special Report on Aviation and the Global Atmosphere" by the Intergovernmental Panel on Climate Change (IPCC), air transportation accounted for 2 percent of all anthropogenic carbon dioxide emissions in 1992 and 13 percent of the fossil fuel used for transportation. In a 10-year period, passenger traffic on scheduled airlines grew by 60 percent; and, air travel was expected to increase by 5 percent for the next 10 to 15 years [IPCC 1999]. With this forecast, aircraft remain an important source of greenhouse gases in coming decades [IPCC 1999]. It was also estimated that in 1992, aircraft were responsible for 3.5 percent of all anthropogenic radiative forcing of the climate and (at the time of the report, were) expected to grow to as much as 12 percent by 2050 [IPCC 1999].

The Committee on Aviation Environmental Protection (CAEP) of the International Civil Aviation Organization (ICAO), an organization of the United Nations (UN), has formed several working groups to address aviation environmental emissions. In addition, the UN Framework Convention on Climate Change (UNFCCC) has promoted a series of multilateral agreements that target values of emissions reductions for the primary industrialized nations [IPCC 1999]. However, prior to SAGE, there was no comprehensive, up-to-date, non-proprietary model to estimate aviation emissions at national or international levels that could be used for evaluating policy, technology and operational alternatives.

Although the degree of projected growth of the air transportation industry may be debated, the unique characteristics of the industry, the influence that they may have upon the environment, and the influence that policies may have upon the industry dictates a clear need for a computer model that analysts can use to predict and evaluate the effects of different policy, technology, and operational scenarios.

Past studies on aircraft emissions have resulted in global inventories of emissions by various organizations including the National Aeronautics and Space Administration (NASA)/Boeing [Baughcum 1996[a,b] and Sutkus 2001], Abatement of Nuisance Caused by Nuisances Caused by Air Transport (ANCAT)/European Commission (EC) 2 group [Gardner 1998], and Deutsche Forschungsanstalt fur Luft- and Raumfahrt (DLR) [Schmitt 1997]. These inventories represent significant accomplishments since they are the first set of "good-quality" global emissions estimates. In this light, SAGE represents the lessons learned from these past studies. Using the best publicly available data and methods, SAGE improves upon these past studies in producing the highest quality emissions inventories to date.

1.2 Objective and Scope

The objective for SAGE is to be an internationally accepted computer model that is based on the best publicly available data and methodologies, and that can be used to estimate the effects on global aircraft fuel burn and emissions from various policy, technology, and operational scenarios. With regard to scope, the model is capable of analyses from a single flight to airport, regional, and global levels of commercial (civil) flights on a worldwide basis.

1.3 Modeling Capabilities

SAGE can generate inventories of fuel burn and emissions of carbon monoxide (CO), unburned hydrocarbons (HC), nitrogen oxides (NOx), carbon dioxide (CO_2), water (H_2O), and sulfur oxides (SOx calculated as sulfur dioxide, SO_2). The three basic inventories generated by SAGE are: (1) four-dimensional (4D) variable world grids currently generated in a standardized 1° latitude by 1° longitude by 1 km altitude format; (2) modal results of each individual flight worldwide; and (3) individual chorded (flight segment) results for each flight worldwide. These outputs and the dynamic modeling environment allow for a comprehensive set of analyses that can be conducted using SAGE.

With the computation modules and the supporting data integrated in a dynamic modeling environment, SAGE provides the capability to model changes to various parameters including those associated with flight schedules, trajectories, aircraft performance, airport capacities and delays, etc. This results in the ability to use SAGE for applications such as quantification of the effects of Communication, Navigation, and Surveillance (CNS)/Air Traffic Management (ATM) initiatives, determining the benefits of Reduced Vertical Separation Minimum (RVSM), investigation of trajectory optimizations, and computing potential emissions benefits from the use of a Continuous Descent Approach (CDA).

1.4 Document Outline

This document provides detailed descriptions of the methods and algorithms used to perform all of the various calculations in generating fuel burn and emissions inventories for a given year. As such, the document is intended to provide technical details of the model rather than serving as a user's guide or as a presentation of output results. The emphasis placed on this document is to provide a completely open and comprehensive coverage of all of the data and methods used within SAGE Version 1.5.

The remainder of this document is organized as follows. Section 2 defines and discusses the various databases that are used by the model. This section serves as background material for the subsequent sections. Section 3 describes the main functional components of the model required to generate a historical inventory of fuel burn and emissions. Section 4 discussed the methods and data required to conduct forecasting in SAGE. Section 5 presents the various outputs of the model including content, format, and uses. Finally, Section 6 provides concluding remarks related to SAGE Version 1.5.

2 DATABASE DESCRIPTIONS

The following sections describe the main input databases used in SAGE. This material serves as background information for the technical discussions in Sections 3 and 4.

2.1 ICAO Emissions Databank

The ICAO emissions databank contains certification data of various jet engines that have entered service. The data was collected from manufacturers through the work carried out by the ICAO Committee on Aviation Environmental Protection (CAEP). The collection stems from emission standards for particulate matter (PM) or "smoke", unburned HC, CO, and NOx [ICAO 1995, ICAO[a] 2004, and QinetiQ 2004]. The primary data used for emissions modeling are the emissions indices (EI) of the various pollutants and the corresponding fuel flows. These and examples of others included within the databank are shown below:

- Engine Name/Type
- Engine Unique Identification (UID) Number
- Type of combustor where more than one type available on an engine.
- Engine type. TF = turbofan, MTF = mixed turbofan
- Bypass ratio
- Engine pressure ratio
- Engine maximum rated thrust, in kilonewtons
- Hydrocarbon emission index (g/kg) at takeoff condition
- Hydrocarbon emission index (g/kg) at climbout condition
- Hydrocarbon emission index (g/kg) at approach condition
- Hydrocarbon emission index (g/kg) at idle condition
- Carbon Monoxide emission index (g/kg) at takeoff condition
- Carbon Monoxide emission index (g/kg) at climbout condition
- Carbon Monoxide emission index (g/kg) at approach condition
- Carbon Monoxide emission index (g/kg) at idle condition
- Oxides of nitrogen emission index (g/kg) at takeoff condition
- Oxides of nitrogen emission index (g/kg) at climbout condition
- Oxides of nitrogen emission index (g/kg) at approach condition
- Oxides of nitrogen emission index (g/kg) at idle condition
- Smoke number at takeoff condition
- Smoke number at climbout condition
- Smoke number at approach condition
- Smoke number at idle condition
- Fuel flow (kg/sec) at takeoff condition
- Fuel flow (kg/sec) at climbout condition
- Fuel flow (kg/sec) at approach condition
- Fuel flow (kg/sec) at idle condition

The power settings for the various modes are defined as:

- Takeoff = 100% power
- Climbout = 85% power
- Approach = 30% power
- Idle = 7% power

2.2 Emissions and Dispersion Modeling System (EDMS) Emissions Data

The Emissions and Dispersion Modeling System (EDMS) is the FAA's required model for assessing the local air quality impacts of aviation sources from proposed FAA airport actions, and is currently listed as an EPA "preferred" model [FAA 2001]. EDMS is used to inventory emissions and model atmospheric dispersion at an airport based upon the emissions of aircraft, auxiliary power units, ground support equipment (GSE), and other sources operating on the airport surface or in nearby vicinity such as on-road passenger vehicles. EDMS was developed jointly by the FAA's Office of Environment and Energy and the US Air Force. Since its original release, EDMS has undergone continuous enhancement to respond to the needs of the user community and to incorporate advances made in the field of local air quality analysis.

The emissions data within the systems tables of EDMS Version 4.2 are in the same format (emissions indices, g/kg) as those within the ICAO emissions databank. These data were obtained from various sources including the US Environmental Protection Agency (USEPA) and various aircraft manufacturers. Because EDMS incorporates the emissions indices from the ICAO emissions databank, the EDMS data is a superset of the ICAO data. With the ICAO jet engine data incorporated within SAGE, the remaining data obtained from EDMS are those corresponding to propeller-driven aircraft (i.e., turboprops and pistons). The EI data for some piston engines from EDMS have been deemed unreliable since they are magnitudes higher than most other engines. Therefore, the current practice within SAGE is not to present results from piston-powered flights.

2.3 Base of Aircraft Data (BADA)

Promulgated by the Eurocontrol Experimental Centre, BADA represents a collection of aircraft performance and operation parameters [EEC 2004]. Currently in Version 3.6, data coverage includes 91 specific aircraft types. But through equivalencies (i.e., matching tables), coverage is promoted to include 295 aircraft types.

The data consists of performance and procedural information categorized into aircraft-specific and global parameters. Included in the data are coefficients that allow calculation of lift and drag forces as well as thrust and fuel flow. As prescribed through the BADA User's Manual, the underlying performance model is based on the use of the Total-Energy Model (TEM) which balances all of the forces acting on an aircraft [EEC 2004] as shown in Equation 1.

$$(T - D)V_{TAS} = mg\frac{dh}{dt} + mV_{TAS}\frac{dV_{TAS}}{dt} \qquad (1)$$

where   T = thrust acting parallel to the aircraft velocity vector (N)
D = aerodynamic drag (N)
m = aircraft mass (kg)
h = altitude (m)

g = gravitational acceleration (9.81 m/s^2)
V_{TAS} = true air speed (m/s)

Coverage of the data/method is for the entire flight regime including all of the different modes, e.g., Landing and Takeoff (LTO) and cruise modes. However, BADA is most accurate for cruise modeling.

Once the various forces are determined, thrust-specific fuel consumption and fuel flow are calculated from the following type of equations [EEC 2004]:

$$\text{Jet:} \quad \eta = C_{f1}\left(1 + \frac{V_{TAS}}{C_{f2}}\right); \quad f = \eta T_k \quad (2)$$

$$\text{Turboprop:} \quad \eta = C_{f1}\left(1 - \frac{V_{TAS}}{C_{f2}}\right)\frac{V_{TAS}}{1000}; \quad f = \eta T_k \quad (3)$$

where η = thrust-specific fuel consumption for takeoff/climbout (kg/min/KN)
C_{f1} = First thrust-specific fuel consumption coefficient (kg/min/kN for jet or kg/min/kN/knot for turboprop)
C_{f2} = Second thrust-specific fuel consumption coefficient (knots)
f = fuel flow (kg/min)
T_k = thrust acting parallel to the aircraft velocity vector (KN)

Equations 2 and 3 apply to the takeoff and climbout modes. Variations of these equations are used for the other modes.

2.4 Integrated Noise Model (INM) and the Society of Automotive Engineers (SAE) Aerospace Information Report (AIR) 1845

The Integrated Noise Model (INM) Version 6.2 represents the FAA's standard methodology for noise modeling and assessments [FAA 2002]. For noise computations, the main component within INM is an internal database of noise vs. power vs. distance (NPD) values, augmented by a database of spectral characteristics. The data requirements are airport conditions, aircraft types, operational parameters, geometry between the observer and flight segment, and noise metric information.

In terms of performance modeling, the main guidance and underlying database/methodology is in large part based on the Society of Automotive Engineers (SAE) Aerospace Information Report (AIR) 1845 [SAE 1986] which provides equations to model aircraft performance such as thrust shown in Equation 4 for jet engines.

$$F_n/\delta = E + F\,v + G_A\,h + G_B\,h^2 + H\,T_C \quad (4)$$

where F_n/δ = corrected net thrust per engine (pounds)
v = equivalent/calibrated airspeed (knots)
h = pressure altitude Mean Seal Level (MSL) (feet)
T_C = temperature at the aircraft (°C)
E, F, G_A, G_B, H = regression coefficients that depend on power state (max-

takeoff or max-climb power) and temperature state (below or above engine breakpoint temperature)

Equation 4 as used in INM actually represents a slight modification to the SAE AIR 1845 equation by using a quadratic for the altitude term ($G_A h + G_B h^2$) rather than the linear estimate ($G h$) prescribed by SAE AIR 1845. Hereinafter, for simplicity, any references made to INM will implicitly include SAE AIR 1845 data and methods as well.

Only LTO modeling up to 3,048 m (10,000 ft) is covered by SAE AIR 1845; the cruise (en-route) mode is not modeled. Performance information from SAE AIR 1845 and procedural information from airlines and manufacturers is used to model flights in INM.

2.5 Airports

The airports database in SAGE represents an aggregation of data from various sources including the FAA, INM 6.2, EDMS 4.2, the Bureau of Transportation Statistics [BTS[b] 2005], Eurocontrol, Manchester Metropolitan University (MMU), and various internet sources. In addition, some data from the commercially available Jeppesen database has been included as well as being used to verify some of the data in these other databases. The aggregated worldwide database within SAGE contains various geographical information including:

- IATA Code
- ICAO Code
- Latitude (deg)
- Longitude (deg)
- Altitude (m)
- Airport Name
- City Name
- Country
- Region

Due to the difficulties with matching International Air Transport Association (IATA) and International Civil Aviation Organization (ICAO) codes for the individual airports, many are simply listed with just one code (either an IATA or ICAO code). As a result, there are over 32,000 records in this database where about one-half to two-thirds are unique airports. This database includes many of the smaller airports, especially in the US.

The altitudes data was generally sparse in the various databases used to form the aggregated SAGE database. Therefore, the altitude data for all airports were systematically obtained from the United States Geological Survey (USGS) GTOPO30 digital elevation model (DEM). 30-arc second (approximately 1 km spacing) elevation data from this model were obtained online [USGS 2005] and implemented within a commercially available Geographic Information System (GIS) tool, ArcGIS [ESRI 2005]. This method provided a consistent and reliable source of data, and was deemed a better approach than trying to fill in the missing ones from various sources which would have been tedious and error-prone.

The Country and Region definitions were also developed using the GIS tool. Airports were assigned to the countries listed within the countries geographical layer based on the latitude and longitude coordinates of the airports. Due to tolerances associated with country boundaries, many airports near country borders had to be manually checked and reassigned so that they were placed into the correct country bins. The list of countries within the GIS tool had to be mapped to a list of countries categorized by regions developed by FAA AEE (Appendix A). Although most of the country names were identical and could be mapped programmatically, many had to be manually mapped due to slight differences in the names (e.g., "United States" versus "United States of America").

2.6 Official Airline Guide (OAG)

The Official Airline Guide (OAG) lists scheduled passenger flights by participating airlines updated monthly. US airlines generally plan out flights quarterly, and may conduct monthly adjustments. Since OAG sells tickets, it represents all US scheduled airlines and the majority of scheduled worldwide airlines in the schedules. The OAG schedules used in SAGE are obtained from the FAA's Office of Aviation Policy and Plans (APO), and lists all trip legs individually [FAA[a] 2005].

The OAG database contains the following data fields:

- LveCode = number representing origin airport
- Leave = origin airport alphabetic code, e.g., BOS
- LveTime = time flight is scheduled to depart origin in local time
- LveGMT = time flight is scheduled to depart origin in Greenwich Meridian Time
- ArrCode = number representing arrival airport
- Arrive = arrival airport alphabetic code, e.g., BOS
- ArrTime = time flight is scheduled to arrive in local time
- ArrGMT = time flight is scheduled to arrive in Greenwich Meridian Time
- Equip = type of aircraft, in code
- FAACarr = abbreviation for air carrier name
- FltNo = flight number
- Freq = 1/0 code showing days of the week that that flight flies that time slot and citypair
- Elapsed = flight time in minutes
- Edate = effective date for beginning this flight schedule
- Ddate = last date this flight will be flown
- ATACarr = carrier name in Air Transport Association Code
- Seats = seats available for sale on this flight
- Deplat = departure airport latitude
- Deplong = departure airport longitude
- Arrlat = arrival airport latitude
- Arrlong = arrival airport longitude
- OAGLve = origin airport in OAG airport code
- OAGArr = destination airport in OAG airport code
- OAGCARR = air carrier company in 2-letter OAG code
- CarrType = commuter or carrier company
- ATAEquip = aircraft type in ATA code
- EqType = J for Jet, T for Turboprop, P for propeller-driven aircraft
- ArrCity = destination city and country/state, spelled out
- CarrName = air carrier company name spelled out
- ArCenter = identifier code for the air traffic facility controlling the destination airspace
- StaMiles = stage miles; number of miles in this leg of flight

- LveCity = origin city and country/state, spelled out
- ArrCntry = destination country or state if the destination is in the US
- LveCntry = origin country or state if the origin is in the US
- YYMM = year and month of the current schedule
- Eday = 0/1 code indicating whether this flight flies on each day of the month given by the schedule
- FPM = number of times (days) this flight is flown between this city-pair at this time slot in a month

OAG does not include unscheduled and charter flights, or general aviation flights including business jets. OAG also does not include military flights or government transfer flights. There are several reasons why the raw OAG schedules may require adjustments to reflect actual traffic. First, the aforementioned unscheduled flights would need to be accounted for either explicitly or through a scaling method. Second, there are known duplicate records in the schedules largely due to the issue of code-sharing. Based on an initial analysis of this data from the FAA, duplicate records were generally found to comprise a negligible number of flights (less than 1%) of the overall schedules. The schedules from the FAA do contain some cargo flights but a significant portion appear to be unscheduled. Therefore, just like any other unscheduled flights, they would need to be accounted for [Appendix L].

2.7 Enhanced Traffic Management System (ETMS)

The Enhanced Traffic Management System (ETMS) is the FAA's electronic recording of flight position and flight plan information used for air traffic management [Volpe 2003]. ETMS captures every flight within coverage of FAA radars, including scheduled, cargo, military, charter, and unscheduled flights. It also captures every flight that files a flight plan, whether or not the aircraft enters radar-controlled airspace. The fields in the flight plan and position data are indicated below:

Flight plan:

- init_char = unused character field
- flight_key = combination of flight UID and flight ID
- flight_date = date of flight operation (either departure date or date of first message)
- flight_uid = flight identifier within ETMS
- flight_id = a "user friendly" identifier for the flight
- first_msg_time = timestamp of the flight's first message
- first_msg_code = internal numeric message type for the first message
- last_msg_time = timestamp of the flight's last message
- last_msg_code = internal numeric message type for the last message
- msg_count = number of messages received for the flight
- schd_dep_airport = scheduled departure airport code (e.g., if from OAG)
- schd_dep_time = scheduled departure time (e.g., if from OAG)
- schd_arr_airport = scheduled arrival airport code (e.g., if from OAG)
- schd_arr_time = scheduled arrival time (e.g., if from OAG)
- plan_dep_airport = planned departure airport code
- plan_dep_center = planned departure Air Traffic Control (ATC) center
- plan_dep_time = planned departure time
- plan_arr_airport = planned arrival airport
- plan_arr_center = planned arrival ATC center
- plan_arr_time = planned arrival time
- act_dep_airport = actual departure airport
- act_dep_center = actual departure ATC center
- act_dep_time = actual departure time
- act_arr_airport = actual arrival airport
- act_arr_center = actual arrival ATC center
- act_arr_time = actual arrival time
- aircraft_type = aircraft code
- number_of_aircraft = number of aircraft in the formation
- max_plan_alt = maximum planned altitude
- cancel_reason = reason for flight cancellation
- physical_class = physical class of the aircraft
- user_class = type of flight (e.g., commercial, cargo, general aviation, etc.)
- weight_class = weight class of the aircraft
- purge_flag = reason the flight was deemed completed (used for debugging)

Flight position:

- init_char = unused character field
- flight_key = combination of flight UID and flight ID
- flight_date = date of flight operation (either departure date or date of first message)
- msg_time = timestamp for the message
- lat = latitude
- lon = longitude
- center = message reporting center
- deltaLat = difference in latitude from the previous message
- deltaLon = difference in longitude from the previous message
- avgAlt = average of the altitudes from the current and previous messages
- altType = type of altitude data (e.g., describes quality)
- avgSpeed = average of the speeds from the current and previous messages
- actMiles = actual nautical miles traveled from the previous message point
- slMiles = straight line nautical miles traveled from the previous message point
- reasonCode = reason for generation of the chord (used for debugging)

The Volpe Center serves as the hub site to for all of the ETMS data receiving a continual flow of data from numerous sources such as Terminal Radar Approach Control Facilities (TRACON), individual airlines, Automated Radar Tracking Systems (ARTS), Air Route Traffic Control Centers (ARTCC), etc. Once the data is received, it is processed at the hub site into a distributed set of databases with access given to various users, e.g., traffic managers at remote sites such as Air Traffic Control System Command Centers (ATCSCC) and Traffic Management Units (TMU). The data are used to monitor traffic, model potential initiatives, issue initiatives, etc.

Coverage of ETMS data includes North America and parts of western Europe (e.g., United Kingdom). The current estimate of flight coverage within ETMS is about 50-60% of total commercial flights worldwide. While ETMS data is generally described as providing radar coverage for the entire flight, the terminal area coverage is limited. In general, most ETMS radar data ends approximately 4000 ft to 10,000 ft above airport elevation.

2.8 BACK World Fleet Registration Database

The world fleet registration database from BACK Aviation contains a listing of worldwide commercial aircraft built since 1940 [BACK 2005]. The database includes all aircraft currently in use and others as indicated by the status categories listed below:

- On order = on the manufacturer's order report, but not yet built
- On option = written agreement between manufacturer and customer that an order may be placed
- Canceled order = a firm order that has been canceled
- Canceled option = an option that has been canceled
- Pre-service, built = built but has not yet been delivered to the customer
- Active, in service = active aircraft that is economically viable
- Temporarily inactive = inactive due to damage, end of lease, or seized by government authority
- In storage = out of service for an extended period of time
- Prematurely destroyed = prematurely destroyed/scrapped due to an accident
- Retired, out of service = retired/scrapped due to the end of the aircraft's useful life

This status information is obtained from various sources including aircraft manufacturer reports and civil aviation registries. In addition to the status, the database includes various aircraft parameters such as:

- AIRCRAFT TYPE
- AIRCRAFT STATUS CODE
- AIRCRAFT SERIAL NUMBER
- PRODUCTION LINE NUMBER
- AIRCRAFT MANUFACTURER
- MANUFACTURE DATE
- DELIVERY DATE
- REG.\TAIL NO.
- ENGINE MANUFACTURER
- ENGINE MODEL
- NUMBER OF ENGINES
- AIRCRAFT NOISE CLASS (STAGE)
- HOURS/CYCLES DATE
- TOTAL HOURS
- TOTAL CYCLES
- EQUIPMENT CATEGORY
- EQUIPMENT TYPE (LAR CODE)
- EQUIPMENT TYPE (LAR CODE)
- EQUIPMENT TYPE (OAG CODE)
- EQUIPMENT TYPE (OAG CODE)
- AIRCRAFT EQUIPMENT MODEL
- FIRST CLASS SEATS
- BUS. CLASS SEATS
- COACH SEATS
- TOTAL SEATS
- OPERATOR CATEGORY
- OPERATOR NAME
- OPERATOR IATA CODE
- OPERATOR ICAO CODE
- OPER. COUNTRY CODE (LAR)
- OPER. COUNTRY CODE (LAR)
- OPER. COUNTRY CODE (OAG)
- OPER. COUNTRY CODE (OAG)
- OPER. REGION CODE
- ACTIVITY DATE
- ACTIVITY
- REMARK
- WING SPAN (M)
- WING AREA (SQ.M)
- OVERALL LENGTH (M)
- OVERALL HEIGHT (M)
- BELLY VOLUME (CU.M)
- MAXIMUM SEATING
- FUEL CAPACITY (L)
- MAX. TAKEOFF WEIGHT (KG)

- MAX. PAYLOAD (KG)
- OPERATING EMPTY WEIGHT (KG)
- MAX. LANDING WEIGHT (KG)
- MAX. ZERO-FUEL WEIGHT (KG)
- RANGE W/MAX. FUEL (KM)
- RANGE W/MAX. PAYLOAD (KM)
- AVERAGE AGE OF AIRCRAFT

The database is typically used by (1) airframe manufacturers to forecast and plan fleet renewal requirements; (2) airline analysts to establish order/delivery dates and to identify aircraft for sale; (3) financial analysts to determine capital requirements, tax allocations, and capital reserves when evaluating owned and leased aircraft; and (4) aircraft component manufacturers to track inventory, orders, and deliveries to forecast demand.

2.9 ICAO's Forecasting and Economics Sub-Group's (FESG) Forecasts

Developed by ICAO CAEP's FESG, the data contains forecasts of number of aircraft, number of aircraft seats, number of flights, stage length, capacity and average seating numbers per aircraft by regions of the world [FESG 2003]. Table 1 shows an example of the first ten years of a typical forecast.

Table 1. Typical 10-Year Projection of Number of Aircraft as a Function of Generic Seat Class

Seats per aircraft	Number of aircraft in the North Atlantic fleet										
	1998	1999	2000	2001	2002	2003	2004	2005	2006	2007	2008
51 - 99	0	0	0	0	0	0	0	0	0	0	0
100 - 150	6	4	4	5	5	5	5	4	5	5	5
151 - 210	201	206	212	216	221	225	230	235	242	249	252
211 - 300	172	178	187	195	205	216	225	234	243	253	261
301 - 400	116	119	122	126	130	133	138	141	145	148	151
401 - 500	15	16	18	20	22	25	28	31	34	38	43
501 - 600	0	0	0	0	0	0	0	1	1	1	3
601 - 650	0	0	0	0	0	0	0	0	0	0	0

Projections were made for the following global regions: North Atlantic, Mid-South Atlantic, Trans-Pacific, Europe-Asia-Pacific, Europe-Africa, Europe-Middle-East, North-South America, Intra-Africa, Intra-Asia-Pacific, Intra Europe, Intra Latin America, Intra Middle East, Other International, Domestic Africa, Domestic Asia-Pacific, Domestic Europe, Domestic Latin America, Domestic Middle East, and Domestic North America.

Because the projections are region-based and overlapping, deriving routing information requires assigning overall traffic to specific city-pairs. While aircraft size serves to segment the forecast, to use this forecast in the emissions model requires forecasting specific airframes and matching them to engine types. The advantage of using these projections is that the member States working within the ICAO CAEP, including the US FAA generally accepted them.

The forecast fleet is also built using, in part, FESG aircraft retirement curves and equations originally developed by Pratt & Whitney. Aircraft retirement modeling is based on the use of four curves:

- An overall original survivor curve for the newer generation aircraft plus the older 2-man flight crew aircraft.
- A new curve for the first generation new bodies.
- A new curve for 727s and the few remaining 707s.
- A new curve for MD-11 passenger aircraft

2.10 FAA's Terminal Area Forecasts (TAF)

Developed by the FAA, the TAF tracks and predicts total annual enplanements, operations, and various FAA workload measures for the airports and control towers that the FAA oversees. The TAF is the official US aviation demand forecast [DOT 1999 and FAA[b] 2005].

The TAF contains historical projected aviation activity data for active airports in the National Plan of Integrated Airport System (NPIAS). The forecasts project activity of four major users of the US air traffic system: air carriers, commuters/air taxi, general aviation, and the military. The FAA uses these forecasts for budgeting and planning. Airport sponsors, state and local aviation authorities, others in aviation industry, and the public may use the data freely.

The data includes both enplanements and operations for each US airport. The FAA not only updates its TAF every year but also improves the forecast methods constantly. The Airport Council International (ACI) sponsors an annual FAA Commercial Aviation Forecast Conference every year. At this conference, new forecasting techniques are proposed and refined. Currently, the FAA derives forecasted operations in the TAF in the following way [DOT 1999]:

- Enplanements are forecasted based on outputs of socioeconomic models such as gross domestic product (GDP) and demographic growth rates with due consideration of originating traffic and connection traffic. Each major airport has its own specific models.
- Forecast load factors are predicted for departures and arrivals at each airport based on the demand, fare yield, and airline costs.
- The average number of seats per aircraft are forecasted for arrivals and departures at the airport.
- Forecasted enplanements are by the forecasted load factors and by the forecasted average number of seats per aircraft to obtain forecasted operations.

In deriving the forecasts, flight delays due to traffic congestion are never explicitly considered. Implicitly, the TAF assumes that airport and ATC capacities will grow to meet the potential demand.

2.11 Airline On-Time Performance Data

Also know as Airline Service Quality Performance (ASQP) data, the US Department of Transportation's (USDOT's) Bureau of Transportation Statistics (BTS) develops the on-time performance database which contains performance data for approximately the 10 largest US carriers [BTS[a] 2005]. Specifically, carriers are required to provide fleet data if they account for 1% or more of the total US scheduled service passenger revenues. The database includes the following data:

- IATA carrier code
- Flight number
- Depart airport

- Arrival airport
- Date of operation
- Day of week
- OAG depart time
- Carrier's computer reservation system (CRS) depart time
- Actual depart time
- OAG arrival time
- CRS arrival time
- Actual arrival time
- OAG-CRS depart delay
- OAG-CRS arrival delay
- CRS elapsed time
- Actual elapsed time
- Actual CRS depart delay
- Actual CRS arrival delay
- Actual CRS elapsed time difference
- Wheels off time
- Wheels on time
- Aircraft tail number
- Taxi out time
- Taxi in time
- Airborne time

2.12 Airport Capacity

As the chief foundation for the delay modeling in SAGE, the capacity data is based on the FAA's annual runway capacity "benchmarking" report of US airports [FAA 2004]. Although the FAA data is based on a two-point definition of arrival-departure Pareto frontiers, the capacity data used in SAGE is based on three points. European airport runway and terminal airspace capacities are based on the airports' own observed and stated capacity limits and noise limitations, as published by Eurocontrol [Eurocontrol 1998].

This capacity data is refined for all weather conditions using analysis and additional data from published airport capacity studies including the European Terminal and En Route Procedure (TERP) manuals. Additional data inputs include the presence or absence of navaids, Instrument Landing System (ILS) capabilities, radar coverage, tower range of view, presence or absence of full runway-length taxiways, high-speed turnoffs, noise restriction zones, and local obstructions.

Represented in the data are 257 airports world wide: 102 US, 122 European, and 33 rest of the world. All together, these 257 airports represent 75% of world commercial travel as defined by the OAG.

2.13 Aircraft Phase-Out Indicators and Fractions

This set of data is used for forecasting within SAGE. The dataset includes a list of world countries with an indicator to specify whether or not the country has adopted the noise Chapter 2 phase-out rule. Also included in the dataset are yearly fractional values which are used to determine the fraction of the fleet within each country that ultimately gets replaced with new aircraft. That is, a country which has adopted the phase-out rule will experience the use of 100% "future aircraft" types for the grown and replacement portion (i.e., due to retirement) of their fleet. The term "future aircraft" is used to reflect near-term state-of-the-art aircraft. If a country has not adopted the phase out rule, it will only experience the use of future aircraft types for a fraction of its grown and replacement portion of their fleet. The remaining fraction (i.e., 1 – fraction) uses the same aircraft type as the existing or surviving aircraft.

This dataset is identical to that used in the FAA's Model for Assessment of Global Exposure due to Noise from Transport Aircraft (MAGENTA), the ICAO CAEP de facto standard for global modeling of aircraft noise. The data was obtained directly from the developers of MAGENTA, Wyle Labs, and its use in SAGE basically mimics the use in MAGENTA [Grandi 2005].

2.14 Aircraft Retirement Parameters

The aircraft retirement parameters are data necessary for aircraft retirement purposes in forecasting. As shown in Table 2, the data is essentially a categorized list of coefficients for survivor curves (polynomial equations) that provide fractions of a fleet population that survived the retirement process.

Table 2. Aircraft Survivor Curve Coefficients

	Aircraft Category 1 7 to 47 years	Aircraft Category 2 7 to 36 years	Aircraft Category 3 12 to 36 years	Aircraft Category 4 5 to 14 years
Constant	0.7912	0.875867	0.277046	0.782491
a	0.0975	0.039574	0.136525	0.080313
b	-0.016835	-0.00352285	-0.0076598	-0.00931738
c	0.0013517	0.0000478103	0.000103682	
d	-0.000053636			
e	-0.00000097731			
f	-6.581E-09			

The dependent variable within the survivor curves is age of the population group, and the aircraft categories are:

- 1: All aircraft except for those corresponding to categories 2-4.
- 2: 1st generation wide body aircraft (A300B4, L1011, DC10, 747-100/200/300).
- 3: B727s and B707s.
- 4: MD-11.

The range of years provided in Table 2 for each category indicate applicability of the survivor curves. For example, the survivor curve for category 1 is first applicable when the average fleet age is 7 years and no aircraft survives after 47 years of service. All of this data was obtained from ICAO CAEP's FESG.

3 METHODOLOGY – HISTORICAL INVENTORY

The fundamental modeling unit in SAGE is a single flight. All data including those related to flight schedules, trajectories, performance, emissions, etc. are set up to support the modeling of a single flight. This simple but powerful concept allows high-fidelity modeling of global inventories of fuel burn and emissions where all flights worldwide for each day of the year are accounted for. And each flight is modeled from gate-to-gate as indicated in Figure 1.

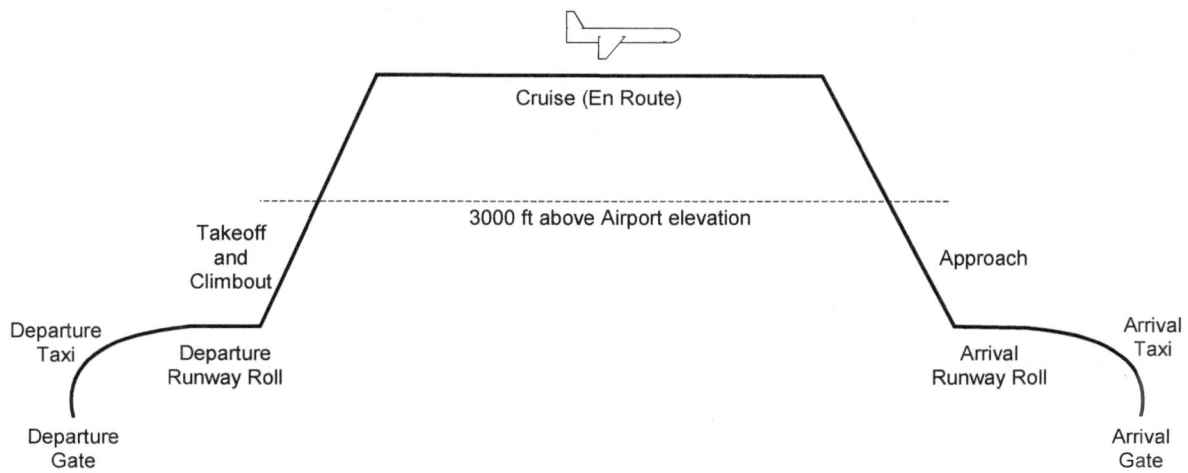

Figure 1. Full Gate-to-Gate Flight Modeling in SAGE

As a result of the full gate-to-gate coverage, all of the modes (i.e., takeoff, climbout, cruise, and approach) are modeled. Although a single flight in SAGE is promoted as the fundamental modeling unit, the actual modeling involves analyzing each individual segment of a flight, hereinafter referred to as a flight chord. Figure 2 exemplifies this by showing the takeoff/climbout trajectories for a single flight represented as a series of finite flight chords.

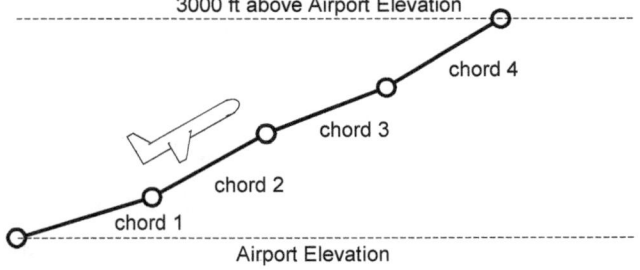

Figure 2. Example Takeoff and Climbout Flight Chords

The flight chords allow the freedom to express the outputs in a variety of different formats (e.g., gridded, per modal totals, etc.) and allows for dynamic aircraft performance modeling in SAGE which results in greater degree of options for scenario modeling. Such modeling is also fundamentally more accurate than those based on aggregated times in mode (TIM) or simple performance lookup tables.

The current worldwide coverage in SAGE includes over 30 million commercial flights per year and accounts for over 200 different aircraft types. Even though substitution aircraft data are necessarily used for some aircraft types, the intention with SAGE is to preserve as much of the specificity of each flight as possible. Therefore, compromises associated with using generic aircraft types and engines are not made.

To accomplish all of this modeling, SAGE includes various modules that process data and perform computations. Figure 3 shows the basic modules as rectangular greyed boxes with the supporting data/methods as elliptical balloons.

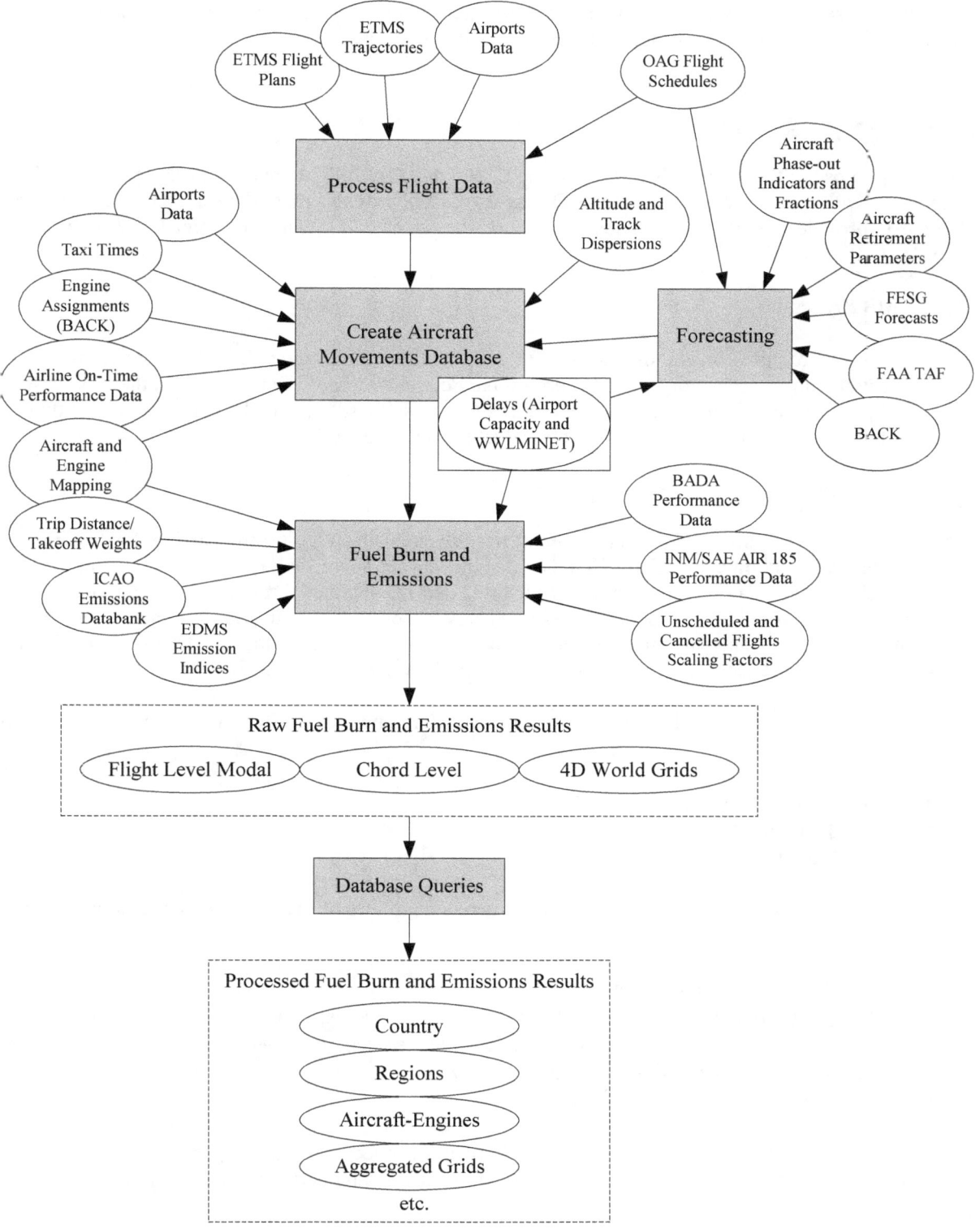

Figure 3. Overview of the Main Modules and Databases in SAGE

The following sections provide descriptions of these modules in the framework of a list of steps required to generate a historical year's inventory of fuel burn and emissions.

3.1 Process Flight Data

The main purpose of this module is to process the ETMS and OAG flight data which serve as the main components within the Aircraft Movements Database. ETMS provides radar data and actual flight plans while OAG provides just flight schedules. Therefore, the use of ETMS data takes precedence over the use of OAG flights.

3.1.1 ETMS Processing

ETMS processing requires several distinct steps: (1) extraction, (2) compression, (3) filtering, (4) cleaning (flight plan and trajectory), and (5) preliminary computations. The results from these processing steps are flight plans and trajectories of actual flights. Because these are actual flights (i.e., rather than scheduled), the data is considered an extremely valuable component of SAGE.

3.1.1.1 Extraction

Extraction involves acquiring ETMS data directly from the hub site at Volpe and parsing the data into separate flight plan and trajectory data. That is, once the various sources report the data to the hub site, the system-processed data (uncompressed binary "orig" data) is obtained through a feed from one of the ETMS servers, and stored on a separate Linux-based server located within the hub site facility. As a result, this data feed is near-real-time. The data is then parsed to obtain separate flight plan and trajectory data which then experiences a compression process described in Section 3.1.1.2.

In parsing the data, an individual flight is first identified when one of the following messages is read from the orig data stream:

- Flight schedule message
- Flight plan message
- Departure message

Then the reading of the flight's data is completed when one or more of the following conditions are encountered:

- An ETMS arrival message was read
- No messages of any kind for 24 hours since the scheduled departure message
- No messages of any kind for 6 hours since the planned departure message
- 6 hours since the last position message (e.g., TZ or TO)
- 24 hours since the last non-position message

As part of this flight identification process, the flight plan (or schedule) data is separated from the position (or trajectory) data into separate data files tied through the use of unique flight identification keys as indicated in Section 2.6.

3.1.1.2 Compression

The compression process occurs as the raw orig ETMS data is parsed into flight plan and trajectory data, and it involves the removal of redundant-type data from the trajectories which are composed of flight chords (or segments). Each trajectory node (or head of each chord) is analyzed through a recursive algorithm to determine whether or not the node provides redundant information. If so, then it is effectively removed, thereby reducing the number of data points required to model a flight. The following criteria are used to judge whether or not a node is removed:

- No change in air traffic control (ATC) center
- No change in altitude type
- No horizontal deviation of more than 2 nm
- No altitude deviation of more than 1000 ft
- No speed deviation of more than 20 kts

All of these criteria must be met in order for the node to be considered redundant and removed. The recursive logic is exemplified by the steps shown in Figure 4 and the following explanation.

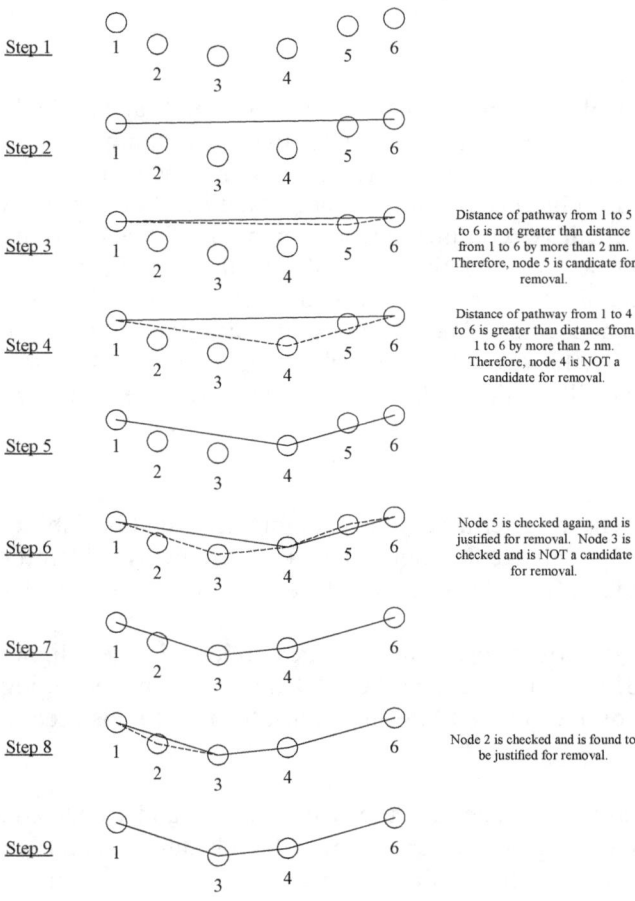

Figure 4. Trajectory Compression

To simplify the explanation, just the horizontal deviation criteria is used as an example. Step 1 in Figure 4 shows an overhead view of a set of 6 fictitious trajectory nodes as they may exist in an ETMS flight. The recursive process begins in Step 2 where the first and last nodes are compared by creating a straight line between the two nodes. Working backwards, temporary lines are created (shown as dotted) in Step 3 between nodes 1 and 5 and between nodes 5 and 6. Assuming that the distance created by these dotted lines are not greater than the solid line's distance by more than 2 nm, node 5 is considered a candidate for removal. The reason why it is not removed now is because there are additional points other than point 1 that can affect the trajectory outcome. Then in Step 4, the next node (4) is analyzed where dotted lines are created from nodes 1 and 4 and from 4 to 6. This time, the 2 nm deviation is exceeded and point 4 is identified as a "keeper" (i.e., not a candidate for removal). Therefore, in Step 5, the lines are made solid. Using the line from 4 to 6 as a base in Step 6, node 5 is again analyzed where dotted lines are drawn from 4 to 5 and from 5 to 6. Since the dotted lines again are not greater than 2 nm, node 5 is now justified for removal. Similarly, the line from nodes 1 to 4 is used as a basis for checking node 3 where lines are created from nodes 1 to 3 and from 3 to 4. The dotted lines are greater than 2 nm, therefore, node 3 is considered a keeper with the dotted lines becoming solid in Step 7. Then in Step 8, node 2 is checked in a similar fashion and found to be justified for immediate removal since there are no nodes preceding it, again, other than node 1. The final trajectory is shown in Step 9. In general, the compression process reduces data size by approximately 30%.

3.1.1.3 Filtering

Filtering involves the removal of flights that have been identified (flagged) in the ETMS database as military or general aviation. Therefore, the remaining flights are mainly commercial but also include flights labeled as "Air Freight/Cargo" and "Air Taxi." Military and general aviation flights may be added in a future version of SAGE. Filtering also involves the flagging of duplicate ETMS flight plan records which may exist due to a repetitive reporting mechanism in ETMS that helps to promote the recording of an accurate flight plan. Due to errors within ETMS, the duplicate flights may not actually be duplicates since they may differ on certain data such as airport codes even though the flight key and flight date may indicate that the flights are duplicates. And because of the difficulty of identifying the correct duplicate, the first flight plan record listed is chosen as the flight to model with all the others labeled as duplicates. In general, duplicate flights will be identical; only a few actually have differing data.

3.1.1.4 Flight Plan Cleaning

Cleaning involves assessments to determine if the data for each flight is viable. And if it is not, then various algorithms are used to salvage the flight. The first step involves identification of flights with missing flight plan information (i.e., airport codes, departure/arrival times, and aircraft types).

For missing airport codes (departure and/or arrival), the duplicate-labeled flight plans are searched to find the missing information. If this fails, then the flight ID and one of the remaining airport codes (if one is available) is used to look for matching ETMS flights; just the flight ID is used if both airport codes are missing.

For missing actual departure/arrival dates and times, the planned and scheduled data are used as substitutes with the planned data given preference over the scheduled data. If neither of these data is available, duplicate flights are searched to find the missing information. If that fails, the flight ID and the departure/arrival airports are used to look for similar flights. The third alternative is to use the flight's first or last chord's date/time message as a reasonable substitute for the missing data. The last alternative is to use the time/date from the flight's first or last message in the flight plan data as a rough approximation for the actual departure/arrival data.

For missing aircraft types, duplicate flights are searched to find the missing information. If the aircraft type cannot be identified from these duplicate flights, similar ETMS flights are searched using the flight ID and departure/arrival airports. If that fails, the same flight ID and departure/arrival airports are used to search through OAG schedules in order to identify the missing aircraft type.

If these preceding steps to fill in the missing data fails, the flight is considered non-viable, and therefore, dropped.

3.1.1.5 Trajectory Cleaning

For the flights that survive the flight plan cleaning process, their trajectories are analyzed to determine whether or not they are viable. The main concern with ETMS trajectories are the quality of the altitude data due to both the magnitude of the errors as well as the shear number of times the errors are encountered. Therefore, most of the emphasis in trajectory cleaning is on the altitude data. In contrast, the horizontal (latitude and longitude) data is considered to be more reliable and they are checked as part of higher-level cleaning efforts. In a future version of SAGE, a more rigorous cleaning process will be implemented to clean the horizontal data.

The first part of the trajectory cleaning process begins with the following high-level checks:

- ETMS total miles traveled is greater then 10,000 nm
- Depart and arrival airports are the same (i.e., Great Circle distance is zero)
- ETMS total miles traveled is less than 50 nm
- ETMS total miles traveled is less then 80 percent of the Great Circle distance
- There are less then 8 ETMS chords

If a flight meets any of these criteria, then it is considered and flagged as "bad" but not dropped as the flight plan information still has the potential to be used. In order to calculate the Great Circle distance, the origin and destination airports from the flight plans are matched to the listings in the airports database in order to obtain the corresponding latitude and longitude coordinates.

In addition to these high-level trajectory checks, faulty altitude data are analyzed by using Rate of Climb (ROC) criteria of 18 m/s and 25 m/s. The 25 m/s ROC is a conservative limit based on state-of-the-art commercial aircraft performance capabilities used to account for bad singular instances of a climb. In contrast, the 18 m/s ROC is based on experiences from analyzing trajectory data and used to account for actual spikes in the data. Both of these criteria are used to smooth the trajectory data.

The smoothing process begins by checking for altitude spikes where two consecutive chords at a time are analyzed to see if they have ROCs >=+18 m/s (climb) and <=-18 m/s (descend), respectively. Once identified, a spike is smoothed by replacing the erroneous point's altitude by the value calculated from the good ROC of the previous chord (i.e., using the available distance and speed at the current point). Partly because this process requires a good previous ROC to begin with, the second ETMS chord is first analyzed using the >=+25 m/s criteria since it is a singular instance. Similar to the spike analysis, the ROC of the first chord is used to calculate a new altitude for any bad ROC occurrences for the second chord (assuming the ROC of the second chord is good). Along with this first chord analysis, the last ETMS chord is also similarly checked. If the ROC for this chord is >=+25 m/s, then the altitude of the last point is set equal to the altitude of the previous point such that the recalculated ROC=0. This was done for simplicity and is based on the assumption that the last chord should have a negative or zero ROC. Although a positive ROC is possible, this assumption was necessary in order to have a negative ROC to be used as the starting point for the individual chord assessments of the latter half of the flight.

The next part of the smoothing process involves analyzing the individual chords to determine if singular instances of bad ROCs occur using the >=+25 m/s and <=-25 m/s criteria. This is accomplished by checking the chords from beginning to middle and then from end to middle. The reason for this is to be able to use the previous chord's good ROC, as the beginning and ending ROCs have already been checked.

The last part of the smoothing process is a secondary check to catch any remaining ROCs that could not be corrected in the previous steps. The chords below 10,000 ft are first nominally categorized as departure and approach chords which are then used to calculate weighted average ROCs for departure and approach. These weighted averages are then used to check the chords in these two modes using the >=+25 m/s criteria for departure and <=-25 m/s for approach. The cruise chords (above 10,000 ft) are also checked using these criteria.

Along with the ROC checks, an additional check for chord points with altitudes less than 0 ft or greater than 45,000 ft is used. If these checks or any of the smoothing process fails for any chord, then the flight is flagged as "bad" similar to the aforementioned high-level checks. The flagged flights are still modeled by treating it similar to an OAG-based flight where trajectories need to be created.

3.1.1.6 Preliminary Computations

Once the flights are cleaned and unique flight keys assigned, some preliminary computations are performed on the trajectory (chorded) data in order to derive the necessary parameters used for fuel burn and emissions modeling. The derived parameters include changes in altitude, speed, and time.

3.1.2 OAG Processing

Processing OAG flight schedules is much simpler than processing ETMS data in part because the data is inherently cleaner and does not include any trajectories. The overall process can be divided into two main parts: expansion and filtering.

3.1.2.1 Expansion

Expansion of the OAG schedules involves the disaggregation of the "Eday" field (see Section 2.5) such that the individual "1"s are replaced by a full record denoting a flight. Therefore, each individual record corresponds to a single flight. There are no issues with multiple trip legs being embedded within a single origin-destination (OD) pair because as indicated in Section 2.5, each OD pair listed in the OAG schedules from FAA represents individual trip legs. Although duplicates due to code-shares exist within the schedules, an analysis of the data has shown that duplicates account for about 1% of all flights listed in the OAG and less than 1% when comparing on a total distance basis (see Appendix B). Therefore, this discrepancy can be ignored.

The expansion process also includes the assignment of unique flight keys and reformatting of the date/time fields to be compatible with the Microsoft SQL database conventions. In addition, unnecessary data are also filtered such as seating and city names.

3.1.2.2 Filtering

The filtering process involves the flagging of helicopter flights and flights less than 50 nm. Helicopters are currently not modeled in SAGE (Appendix C). Flights less than 50 nm are also not modeled because the trajectories necessary to model them are difficult to accurately predict. Because these flights represent a small fraction of total flights and distance traveled, they can effectively be ignored.

3.2 Create Aircraft Movements Database

The creation of the aircraft movements database represents an aggregation of ETMS, OAG, and various supporting data. Although this is shown as a separate process in Figure 3 for clarity, it can actually be thought of as a continuance of the flight data processing. The purpose of both sections is to ultimately create a processed movements database that can be used to model fuel burn and emissions. The following sections describe each data component and the associated methods used to create the movements database.

3.2.1 Development of Aircraft and Engine Mapping Tables

Flights in SAGE are modeled through the use of aircraft codes specified in the ETMS and OAG flight plans and schedules, respectively. These codes are mapped to the aircraft listings within the BADA and INM performance databases. In many cases, the mappings are perfect (or nearly so) as shown by the following example:

- OAG "B734"
- BADA "B734"
- INM "737400"

In other cases, the mappings require substitutions because an exact match to the ETMS/OAG aircraft type may not exist in the BADA/INM databases:

- OAG "A341"
- BADA "A343"
- INM "A340"

In ETMS and OAG, there are approximately 6000 and 230 unique aircraft codes, respectively, from each of these data sources, and they all fall into one of the following categories:

- Good code based on ICAO or IATA
- Bad code but identifiable
- Bad code Unidentifiable

Essentially, any codes that are not readily identified as an ICAO or IATA standard code are termed "bad." Much of these bad codes are unidentifiable and therefore, the associated flight(s) cannot be modeled. However, based on rankings of operations by aircraft code, approximately 95% of the ETMS flights and all of the OAG flights are salvaged. That is, even though the majority of the ETMS aircraft codes are not identified, the approximately 500-600 codes that are identified account for 95% of those flights. Therefore, about 5% of ETMS flights cannot be modeled due to unidentifiable aircraft codes. Although it is possible to investigate the unidentified ETMS aircraft codes further, the resources required to investigate the remaining codes are not justified to salvage the small number of flights, especially when most of them have OAG equivalents (i.e., they are modeled as OAG flights rather than ETMS flights).

The mappings are based on engineering judgment and investigating various aircraft performance-related sources including Jane's All the World's Aircraft [Jane's 2005] and www.airliners.net [Lundgren 2005].

Mappings are conducted based on preferences given to airframe families, comparing thrust/power ratings, aircraft size/weight, etc. Appendix D shows the aircraft mapping table used in SAGE.

In addition to the mappings used to model aircraft performance, aircraft mappings are also necessary for the engine assignment process (see Section 3.2.2). Since engines are assigned based on airline and aircraft categories developed from analyzing the BACK world fleet, mapping is necessary to match BACK codes to ETMS and OAG codes. This mapping is fortunately easier because BACK provides additional information that can be used to more readily identify aircraft types. These aircraft mappings are provided in Appendix E.

In SAGE, the aircraft types are used to model aircraft performance and fuel burn. Although the aircraft types specified in BADA and INM have default engines, the airframes are promoted in SAGE to model aircraft performance rather than the aircraft-engine combination. To model emissions, engines are assigned to the flights (see Section 3.2.2), and their corresponding ICAO and EDMS emissions data are used. These assigned engines are all based on listings as found in the BACK world fleet database. As such, the engine names (or codes) are not identical to the ICAO or EDMS engine names. Hence, mapping tables are also required.

Appendix F shows the engine mapping table used in SAGE. Similar to the airframes, the engines are mapped with preferences given to the same families and by comparing thrust/power ratings, as well as the type and size of aircraft the engines are typically used. Unlike the aircraft code mappings, all BACK engines have been mapped to either an ICAO or EDMS code.

3.2.2 Conduct Engine Assignments

The ETMS and OAG flight plan and schedule data do not provide engine listings. Therefore, they are assigned based on one of three methods. The first and preferred method is through an exact assignment by identifying the tail number of the aircraft from the BTS Airline On-Time Performance data through matchings of flight ID numbers and aircraft type. Once this is accomplished, the tail number is looked-up in the BACK database and the exact engine is assigned. Since the BTS data only nominally covers the top 10 US airlines, many flights cannot be assigned exact engines.

A difficulty with this first approach is that not all tail numbers from BTS will match with those available in the BACK database. This is due to encoding differences between the different systems. As a result, some supplementary identifications and matchings of tail numbers are conducted using JP Fleets data [JP Fleets 2004]. JP Fleets provides additional airline-specific registration-type numbers.

The second method is to assign engines based on popularity within the world fleet. The BACK world fleet database is used to developed distributions of engine counts based on airline and aircraft categories as exemplified in Table 3.

Table 3. Example Engine Distribution

Airline	BACK Aircraft	BACK Engine	Fraction
XYZ	A3XX	PW123	0.43
		JT123	0.57
	B7XX	PW123	0.21
		JT123	0.65
		CFM123	0.14

The fractions always add up to 1 within each airline-aircraft category, and represent the distribution of engines within that category. As indicated in Section 3.2.1, the aircraft code mapping table in Appendix E is used to develop the distributions.

To assign an engine to a flight, the airline and aircraft codes are first matched to those of the flight, and then an engine is randomly picked from the distribution within the airline-aircraft category. When the overall counts of engines assigned to flights analyzed, they will closely match the distributions developed from BACK. The assumption with this method is that popularity within a fleet resembles frequency of usage.

The third method involves the use of default engines. In some cases, the airline codes or a combination of the airline and aircraft codes will not match any from BACK. In that case, the default engines provide the only recourse to assigning engines. These default engines are generally the most popular engines found in the world fleet (see Appendix D). The use of these default engines is similar to the exact engine assignments in that they do not account for the majority of engine assignments. Most of the engines are assigned based on the distributions from the second method as shown by the approximate percentages of flights using the different methods obtained from analyzing a month's worth of flights in 2000:

First Method (BTS tail number match): 14%
Second Method (BACK distribution): 77%
Third Method (Default engine): 9%

3.2.3 Develop Cruise Altitudes and Trajectories

In order to model flights that do not have trajectories (e.g., OAG flights and ETMS flights with bad trajectory data), a trajectory generation method was developed by analyzing a large set of ETMS flights. Specifically, two months (May and October) in 2000 and 2003 were statistically analyzed to develop distributions of cruise altitudes and horizontal tracks. The resulting distributions used in SAGE Version 1.5 are an improvement over the previous distributions used in Version 1.0 [Michot 2004]. Conceptually, the distributions of altitudes and tracks provide a "dispersion" effect (e.g., dispersed around the Great Circle) as shown in Figure 5.

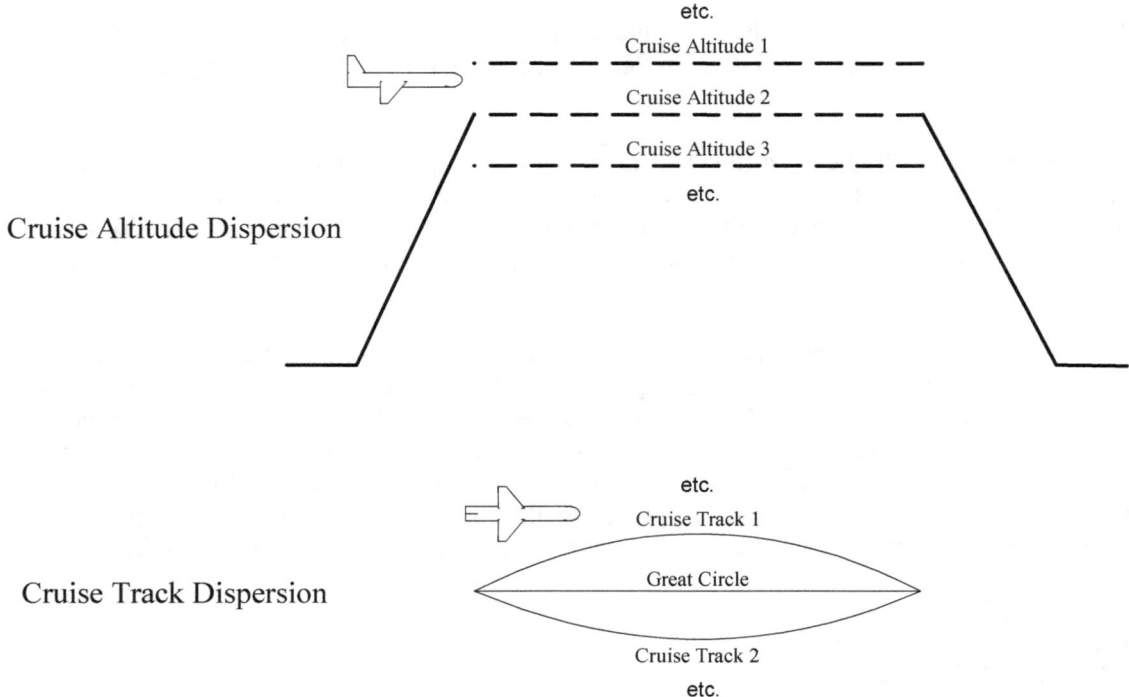

Figure 5. Cruise Altitude and Track Dispersions.

The distribution of cruise altitudes and cruise tracks are presented in appendices G and H, respectively. Only a portion of the cruise track distributions (those falling into the 200 to 250 nm flight distance category) are shown due to the large quantity of the data. The distributions are categorized into jet and turboprop types, and based on flight distance. The track distributions were developed using offsets from the Great Circle route. That is, when a dispersed track is picked from the distribution, it is defined by a set of perpendicular offsets from the Great Circle spaced equally along the Great Circle starting from 20% from the beginning and finishing 80% from the end. The increments between these two points are in 10% ranges.

3.2.4 Develop Taxi Times

Taxi time as defined in SAGE is just the time an aircraft takes to move from the gate to the runway without any impediments or idle waiting just before it begins its takeoff roll. Therefore, SAGE taxi times do not include delays which are modeled separately (Section 3.2.5). The source of taxi data in SAGE is the Airline On-Time Performance data from BTS. Since the data from BTS includes delays, the individual listings of flights are processed to develop percentiles of taxi data for each of the airports covered. The taxi data for the 15th percentile are used to represent "pure" taxi times. The 15th percentile was chosen as a reasonable compromise between choosing a low-enough percentile to ensure that no delays were included and also preventing any unreasonably low values (e.g., reported airline mistakes) from being used.

The resulting average departure and arrival airport-specific taxi times are applied to each flight based on the departure and arrival airports. If an airport is not covered by the BTS data, then the average of all

BTS-covered airports are used. A nominal flight chord is created to model the time duration of each taxi time.

3.2.5 Calculate Delays

Delays are modeled in SAGE through a sub-model called, WWLMINET. Details of this methodology and supporting data are presented in Appendix I. WWLMINET is a "worldwide" version of LMINET, which is a queuing model that predicts hourly airport ground and approach airborne delays [NASA 1998]. It does this by running a flight demand through a network of queues and determining the delays associated with serving that demand level. The WWLMINET network currently contains 102 US airports, 122 European airports, and 33 other airports (i.e., outside of US and Europe). Together these 257 airports represent approximately 75% of global commercial air traffic as defined by the OAG schedules. Airports not included in this network are assumed to have no delays.

Most of the information within WWLMINET including airport capacity and average weather visibility data are held constant when modeling delays to develop historical inventories of fuel burn and emissions. The reason for this is due to the difficulty of obtaining yearly variations of such data. But this is considered reasonable as the capacities and average visibility data will not likely changed drastically from one year to another. And even though the data is kept constant for historical inventory development purposes, the data can be modified to reflect changes at an airport. Unlike the capacity and visibility data, the demand information fed into WWLMINET is different for each year.

The demand data takes the form of a week's worth of OAG flight schedules: May 29 to June 4. This specific week was chosen from an analysis of comparing fuel burn and emissions data for all 52 weeks in each of the four inventories (2000-2003) from SAGE Version 1.0 (see Appendix J). Modeling a week was considered to be sufficient in providing reasonable estimates of delays as it considered the daily traffic variations and the week itself was chosen from a statistical analysis.

The data generated from WWLMINET are hourly departure and arrival ground-based delays and arrival airborne delays (i.e., due to holdings). Similar to taxi times, nominal flight chords are created to model the time durations of the ground-based delays. To model the time duration associated with an arrival airborne delay, a nominal flight chord positioned geometrically between the arrival airport elevation and cruise altitude is created. Airborne delays are modeled when only schedule (i.e., from OAG) or flight plan (i.e., from ETMS) data are available. If the trajectory data for an ETMS flight is viable, then it is assumed that the trajectory data already includes airborne delay. In contrast, ground-based delays are modeled for both OAG and ETMS flights. As indicated in Section 2.6, ETMS radar is very limited in the terminal areas and would generally not provide ground movement details.

3.2.6 Finalize Movements Database

The finalization process involves aggregating the various information into one database (multiple tables) that can be used efficiently during the fuel burn and emissions calculations. After all of the processing and assignment of equipment and trajectories, the fields for the final movements data are shown below:

ETMS flight plan:

- flight_key = unique SAGE flight key
- etms_uid = same as original ETMS flight key
- flight_date = flight date
- first_msg_time = timestamp of the flight's first message

- last_msg_time = timestamp of the flight's last message
- flight_id = a "user friendly identifier for the flight
- raw_msg_count = count of messages received
- schd_dep_airport = scheduled departure airport code
- dep_airport = actual departure airport code
- schd_arr_airport = scheduled arrival airport code
- arr_airport = actual arrival airport code
- schd_dep_time = scheduled departure time
- dep_time = actual departure time
- schd_arr_time = scheduled arrival time
- arr_time = actual arrival time
- physical_class = physical class of the aircraft
- user_class = type of flight
- weight_class = weight class of the aircraft
- purge_flag = reason the flight was deemed completed
- status_flag = flight quality check flag
- time_flag = time quality check flag
- cancel_reason = reason for flight cancellation
- aircraft_type = aircraft code
- tail_number = aircraft tail number
- engine_model = assigned engine
- taxi_out_time = assigned taxi-out time
- taxi_in_time = assignend taxi-in time
- cruise_altitude = assigned cruise altitude (if radar data not viable) or maximum altitude reached (if radar data is viable)
- track_no = track dispersion number

ETMS Flight Chord:

- flight_key = unique SAGE flight key
- msg_time = timestamp for the message
- latitude = latitude at tail of the chord
- longitude = longitude at tail of the chord
- altitude = altitude at tail of the chord
- h_i = average altitude of the chord
- delta_alt = change in altitude for the chord
- v_i = average speed for the chord
- delta_v = change in speed for the chord
- delta_t = change in time for the chord
- act_miles = distance of chord

Streamlined OAG schedule:

- flight_key = unique SAGE flight key
- oag_uid = unique OAG flight ID
- flight_date = flight date
- status_flag = OAG flight quality check flag
- cruise_altitude = assigned cruise altitude

- track_no = assigned track dispersion number
- engine_model = assigned engine
- etms_flight_key = matching ETMS flight key
- etms_flight_status = ETMS flight quality check flag
- tail_number = aircraft tail number
- taxi_out_time = assigned taxi-out time
- taxi_in_time = assigned taxi-in time

The OAG and ETMS flight plan and schedule data are stored in separate tables for clarity and efficiency in modeling the two different types of data. Unlike taxi times, delays are not included in the movements database; they are obtained dynamically from an hourly lookup table during fuel burn and emissions calculations.

The use of these finalized OAG and ETMS flights to model fuel burn and emissions involves identifying the OAG flights for which ETMS matching flights exist. Rather than post-processing the flights to remove matching OAG flights, the "removal" is conducted prior to their use. That is, during fuel burn and emissions calculations, the ETMS-matched OAG flights are excluded from being run

3.3 Preparation of Data for Fuel Burn and Emissions Calculations

Once the movements database has been created, the next step is to prepare the data needed to support the fuel burn and emissions calculations. As indicated in Figure 3, the fuel burn and emissions module (FBE module) requires the following sets of data: (1) takeoff weights, (2) aircraft performance data, (3) emissions data, and (4) unscheduled and canceled flights scaling factors.

3.3.1 Trip Distance/Takeoff Weights

Takeoff gross weight (TOGW) is modeled through the use of weights associated with stage lengths from INM. An example set of takeoff weights and the corresponding stage lengths are shown for the B747-200 aircraft in Table 4.

Table 4. Example Table of Aircraft Takeoff Weights as a Function of Trip Distance

Aircraft ID	Stage	Trip Distance (nm)	Takeoff Weight (lbs)
B747-200	1	Dist < 500	525000
B747-200	2	$500 \leq$ Dist < 1000	545000
B747-200	3	$1000 \leq$ Dist < 1500	565000
B747-200	4	$1500 \leq$ Dist < 2500	610000
B747-200	5	$2500 \leq$ Dist < 3500	665000
B747-200	6	$3500 \leq$ Dist < 4500	725000
B747-200	7	Dist $\geq$ 4500	775000

As indicated, the takeoff weights are essentially a function of trip distance. The number of stages available for an aircraft types will depend on its flight range capabilities and the granularity of the trip distance ranges exemplified in Table 4. For instance, smaller (shorter range) aircraft types may only have one or two stages whereas larger aircraft like the B747-200 may have much more stages.

To account for the practice of fuel tankering used by many airlines to take advantage of lower fuel prices in certain regions, the takeoff weights for all flights were increased systematically by two stage lengths. This method follows on the precedent set by various noise studies that have been conducted using INM to more accurately model takeoff weights. This method was thought to be reasonable since the competitiveness of the airlines would drive them to fuel tanker whenever possible.

In increasing the takeoff weights by two stage categories, the situation arises when there are no more stage categories to move up to. In those cases, the difference between the current and previous stage weights is used to increase the current stage weight.

3.3.2 Aircraft Performance Data

Aircraft performance data is obtained from two sources: BADA and INM. BADA provides aircraft performance data for cruise while INM provides performance for the LTO modes. BADA's speed schedule is used for LTO for consistency and BADA's fuel flow coefficients are used for all modes since INM currently does not model fuel flow. Although BADA can be used to model all modes, the INM data and methods were implemented for LTO modeling because INM has been in existence much longer than BADA, and has not only been extensively validated, but is internationally recognized and accepted as well [Flathers 1982].

The performance data are extracted from the sources and reformatted into new tables as shown in Appendix K. Although reformatted, the actual data value and units are kept identical to those form the sources. All of the data are used as prescribed by the BADA and INM equations with the exception of modifications to the BADA drag coefficients (parasitic and induced). The reasons for the modifications are due to the fact that BADA lacks non-cruise modal coefficients for many aircraft types. That is, data for cruise are readily available, but the others are lacking. The guidance provided in the BADA user's guide [EEC2004] prescribes the use of the cruise coefficients when the data for the other modes are missing.

A better solution to the use of the cruise coefficients was to develop scaling factors that could be used to modify the cruise coefficients to be more representative of the other modes. These factors were developed by dividing the average coefficients for the other modes by the average cruise coefficients for those aircraft where data was available. The factors are presented in Table 5.

Table 5. Modal Drag Coefficient Scaling Factors

Aircraft Category	Initial Climb		Takeoff		Approach		Landing		Landing Gear
	Parasitic	Induced	Parasitic	Induced	Parasitic	Induced	Parasitic	Induced	Parasitic
Jets	1.143	1.071	1.476	1.010	1.957	0.992	3.601	0.932	1.037
Turboprops	1.000	1.000	1.220	0.948	1.279	0.940	1.828	0.916	0.496

Further categorizations into aircraft size categories were explored (e.g., large, medium, and small), but they did not appear to produce significant differences. Therefore, just the overall jet and turboprop categories were used. Application of the factors involves simply multiplying by the cruise coefficient. For example, to obtain the takeoff parasitic drag coefficient (AD_Cdo_IC from Appendix K) for a jet aircraft, multiply the cruise parasitic drag coefficient (AD_Cdo_CR from Appendix K) of that aircraft by the corresponding scale factor (1.476 from Table 5).

3.3.3 Emissions Data

Emissions data in SAGE is obtained from two sources: ICAO emissions databank and EDMS. The emissions data from these sources are processed to remove unused data parameters, thereby resulting in the four modal sets of fuel flow rates and emissions indices for CO, HC, and NOx. In addition, the engine name and UID are also preserved. Examples of unused data include pressure ratios, characteristic emissions, engine status, etc.

To facilitate the use of the data in Boeing's Fuel Flow Method 2 (BFFM2) [Baughcum 1996], the log values of the fuel flow rates and emissions indices are calculated and stored in the emissions data tables along with the line fitting parameters associated with the BFFM2 methodology. Specifically, these log values and additional parameters are:

- LOGRWf_TO = Log value of adjusted (for installation effects) takeoff fuel flow rate
- LOGRWf_CO = Log value of adjusted (for installation effects) climbout fuel flow rate
- LOGRWf_AP = Log value of adjusted (for installation effects) approach fuel flow rate
- LOGRWf_ID = Log value of adjusted (for installation effects) idle fuel flow rate
- LOGCO_REI_TO = Log value of takeoff EI for CO
- LOGCO_REI_CO = Log value of climbout EI for CO
- LOGCO_REI_AP = Log value of approach EI for CO
- LOGCO_REI_ID = Log value of idle EI for CO
- LOGHC_REI_TO = Log value of takeoff EI for HC
- LOGHC_REI_CO = Log value of climbout EI for HC
- LOGHC_REI_AP = Log value of approach EI for HC
- LOGHC_REI_ID = Log value of idle EI for HC
- LOGNOx_REI_TO = Log value of takeoff EI for NOx
- LOGNOx_REI_CO = Log value of climbout EI for NOx
- LOGNOx_REI_AP = Log value of approach EI for NOx
- LOGNOx_REI_ID = Log value of idle EI for NOx
- LOGCO_m = Slope of line formed by log values of idle and approach CO points
- LOGCO_b = Y-intercept of line formed by log values of idle and approach CO points
- LOGCO_RWfCross = Log fuel flow rate at which the two log-log lines intersect for CO
- LOGCO_REIConstant = Log EI value that defines the horizontal line for CO
- LOGHC_m = Slope of line formed by log values of idle and approach HC points
- LOGHC_b = Y-intercept of line formed by log values of idle and approach HC points
- LOGHC_RwfCross = Log fuel flow rate at which the two log-log lines intersect for HC
- LOGHC_REIConstant = Log EI value that defines the horizontal line for HC
- LOGNOx_m = Slope of line fitted through log values of all four points for NOx
- LOGNOx_b = Y-intercept of line fitted through log values of all four points for NOx

3.3.4 Unscheduled and Canceled Flights Scaling Factors

Due to the use of OAG flight schedules for areas outside of ETMS coverage, unscheduled and cancellation of flights cannot be directly modeled. Instead, their effects are indirectly accounted for through the use of scaling factors that generally have the effect of increasing the number of flights. As explained in Appendix L, a comparison analysis of large sets of ETMS and OAG flights were conducted to determine relationships between the number of operations (departures) at an airport and the number of unscheduled and canceled flights as indicated by equations 5 and 6.

$$UF = 12.43653 + (0.09164)SF - (0.000003)SF^2 \tag{5}$$
$$CF = 0.1728847 + (0.024352)SF + (0.000001)SF^2 \tag{6}$$

where UF = Unscheduled flights in a week
CF = Canceled flights in a week
SF = Scheduled flights in a week

These equations are used to predict the number of actual flights at an airport by adding and subtracting the unscheduled and canceled flights, respectively, from the scheduled flights as shown in Equation 7. The resultant actual flights are then divided by the scheduled flights in order to obtain a scaling factor for that airport as indicated by Equation 8.

$$AF = SF + UF - CF \tag{7}$$
$$SCF = AF/SF \tag{8}$$

where AF = Actual flights in a week
SCF = Scaling Factor used to account for unscheduled and canceled flights at an airport

The scaling factor is then used systematically for all flights originating (departing) from that airport. That is, all flight results (e.g., fuel burn, emissions, distance, etc.), associated with flights origination from that airport are multiplied by the scaling factor. Because the scaling factor is a function of OAG scheduled flights, the strength of the factor is directly related to the number of OAG flights modeled at an airport. Therefore, areas where ETMS coverage is abundant will generally have scale factors closer to unity than areas predominantly covered by OAG flights. An assumption with this method is that when the flights in a region are modeled using both OAG and ETMS flights, the ETMS flights include a random mix of scheduled, unscheduled, and canceled flights rather than being biased toward covering, for example, only unscheduled flights. Such a bias is highly unlikely as there are no rules or mandates within the ETMS reporting tools that would produce such results. That is, the limited coverage of ETMS in certain areas is simply due to a lack of reporting mechanisms currently in place in that area rather than a concerted effort to only report a few unscheduled flights. And because of the nature of these same reporting mechanisms, accidental (or coincidental) biases are also highly unlikely to occur.

The use of this scaling methodology requires the development of airport-specific scaling factors. This is accomplished programmatically by feeding in an average week's worth of OAG scheduled flights as modeled in SAGE for each airport into equations 5 and 6 to ultimately obtain the scaling factors from Equation 8. The average week is obtained by dividing total yearly flights by 52. Therefore, a tabular listing of airports and the corresponding scaling factors is created for each year.

In developing equations 5 and 6, several worldwide airports were identified as being outliers. Many of these were identified as being predominantly used by cargo carriers (e.g., FedEx, UPS, etc.) or at least being strongly influenced by them. As a result, they are not modeled using equations 5 and 6, but rather using the separate airport-specific factors indicated in Table 6.

Table 6. Scaling Factors for Outlier Airports

Airport Code	Airport Name	Scaling Factor
MEM	Memphis, TN	1.222
IND	Indianapolis, IN	1.144

SDF	Louisville, KY	1.385
ANC	Anchorage, AK	1.550
MIA	Miami, FL	1.147
TJSJ	San Juan, PR	1.127
MMUN	Cancun, Mexico	1.317
CYYZ	Toronto, Canada	1.086
CYYC	Calgary, Canada	1.110
CYVR	Vancouver, Canada	1.077
CYWG	Winnipeg, Canada	1.131
EGSS	Stansted, London	1.131
LFPG	Charles-De-Gaulle, Paris	1.094
EDDK	Cologne-Bonn, Germany	1.274
EDDL	Dusseldorf, Germany	1.076
EGNX	East Midlands, UK	1.419
EDDF	Frankfurt, Germany	1.064
EHAM	Amsterdam, Netherlands	1.094
EBBR	Brussels, Belgium	1.203
LIMC	Milan, Italy	1.092

3.4 Calculate Fuel Burn and Emissions

The FBE module is a dynamic performance model used to generate the three basic SAGE outputs (flight-level, chord-level, and grids) of fuel burn and emissions. Because the module is dynamic, it allows assessments of various scenarios involving changes to the aircraft performance and emissions data. The key components of the module are: (1) atmospheric data, (2) aircraft performance modeling, (3) fuel flow and fuel burn computations, (4) emissions modeling, and (5) additional data checks. A more detailed pseudo-code description of the module can be found in Appendix M and in Lee 2005.

3.4.1 Atmospheric data

Atmospheric data such as temperature and pressure are based on the use of the International Standard Atmosphere (ISA) where temperature and pressure at sea-level are defined as 288.15 K (59°F) and 101325 Pa (1 atm), respectively. Temperature variations with altitude below the tropopause are calculated according to the standard lapse rate as exemplified by Equation 9.

$$T = T_o - (0.0065 \text{ K/m})h \tag{9}$$

where T = temperature at altitude h (K)
T_o = temperature at sea-level (K)
h = altitude above sea-level (m)

At or above the tropopause, temperature is held constant at 216.65 K, and under ISA conditions, the tropopause is defined as 11,000 m. Pressure variations with altitude below or at the tropopause are calculated according to Equation 10.

$$P = P_o(T / T_o)^{5.2579} \tag{10}$$

where P = pressure at altitude h and temperature T (Pa)

P_o = pressure at sea-level (Pa)

Above the tropopause, pressure is calculated using Equation 11.

$$P_{above} = P_{trop}\exp[-(g / RT_{trop})(h-h_{trop})] \qquad (11)$$

where P_{above} = pressure at altitude h above the tropopause
P_{trop} = pressure at the tropopause = 22,619 Pa
g = gravitational constant = 9.81 m/s^2
R = gas constant for air = 286.9 J/kg-K
T_{trop} = temperature at the tropopause = 216.65 K
H_{trop} = tropopause altitude = 11,000 m

Future versions of SAGE may incorporate actual weather data either as a complete replacement of the lapse rate-generated data or in combination where the lapse rates could be used to fill in data where actual weather data is deficient.

In addition to temperature and pressure, the relative humidity and specific heat ratio are assumed to be constant at 60% and 1.4, respectively. Also, winds are currently not modeled in SAGE, but placeholders exist for their inclusion when the aforementioned actual weather data is implemented.

3.4.2 Aircraft Performance Modeling

Aircraft performance modeling requires several steps: (1) Create takeoff/climbout and approach trajectories, (2) determine aircraft speed schedule, (3) calculate lift and drag, and (4) calculate thrust and rate of climb.

3.4.2.1 Calculate Takeoff/Climbout and Approach Trajectories

The flight modeling process begins with the generation of takeoff/climbout and arrival trajectory chords. Placeholder chords at the altitudes shown in Figure 6 are first created.

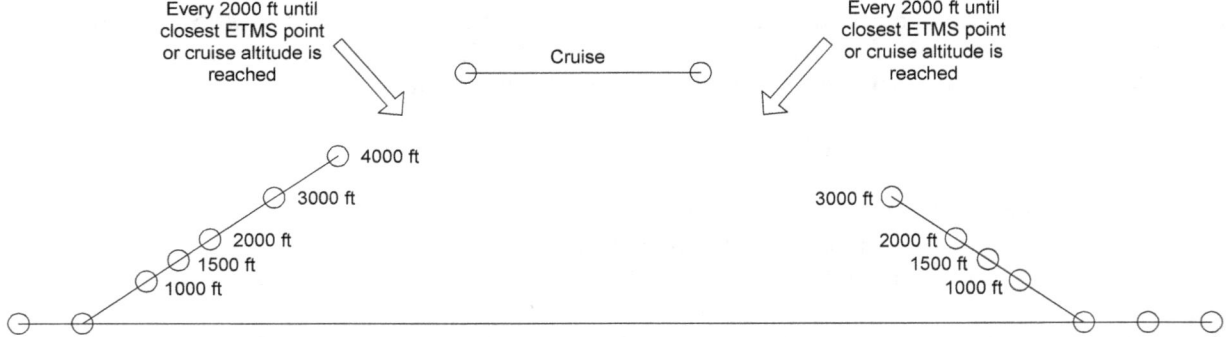

Figure 6. Takeoff/Climbout and Approach Trajectories

As shown in the figure, 1000 ft altitude steps are used until 4000 ft for takeoff/climbout and 3000 ft for approach. Increments of 2000 ft are then used to reach either the closest ETMS point or the assigned cruise altitude for an OAG flight. This method is used for both OAG and ETMS flights because ETMS does not provide the fidelity or coverage in the terminal areas. Although it is possible that ETMS trajectory points below 3000 ft may exist (rarely), they are not used for consistency of the method.

All of these altitudes are fixed except for 1500 ft, 2000 ft, and 4000 ft due to the fact that they represent the tail ends of chords that experience acceleration movements. The algorithm used to model these movements requires a convergence of starting and calculated altitudes in order to determine the final altitude an aircraft will reach. Therefore, the algorithm will start with the 1500 ft, 2000 ft, and 4000 ft altitudes and will iterate to converge upon the final altitudes. Appendix M provides details on this algorithm in Step 7 of the Main Module Section.

as calculated through an INM performance equation (see Appendix M) and through the use of the total energy model (see Equation 1). The convergence is conducted through an iteration of altitude modifications.

The generation of these placeholder altitudes within the fuel burn and emissions module (FBE module) rather than as part of the movements data helps to make the module more dynamic in terms of modeling a full flight. In addition to the airborne chords, one takeoff runway roll chord and two arrival runway roll chords are also created. The takeoff runway roll chord is modeled dynamically along with the airborne chords, but the two arrival roll chords represent fixed data (specific to each aircraft type) from INM based on standard day sea-level conditions. That is, the aircraft-specific chord distances, speeds and thrust are fixed. With the altitudes known for these takeoff/climbout and approach chords, supporting data such as changes in altitude and all of the necessary atmospheric data (Section 3.4.1) are also calculated.

3.4.2.2 Determine Aircraft Speed Schedule

Aircraft speeds are obtained from the BADA airline procedure models which are essentially a set of speeds specified for each altitude band. The specifications of the speeds are based on various BADA speed parameters provided in Appendix K. This speed data is incorporated under the INM procedural framework within SAGE. That is, the INM procedures corresponding to the 1000 ft, 1500 ft, 2000 ft, etc. altitudes shown in Figure 5 specify what BADA speeds are used at those altitudes.

A couple of example specifications for takeoff/climbout from the BADA User Manual [EEC 2004] are reproduced as equations 12 and 13.

$$\text{From 0 to 1,499 ft:} \quad V_{CAS} = C_{Vmin}(V_{stall})_{TO} + Vd_{CL,1} \quad (12)$$
$$\text{From 10,000 ft to transition:} \quad V_{CAS} = V_{cl,2} \quad (13)$$

where V_{CAS} = calibrated air speed (CAS) for the altitude band specified (knots)
C_{Vmin} = minimum speed coefficient
$(V_{stall})_{TO}$ = takeoff stall CAS (knots)
$Vd_{CL,1}$ = speed schedule CAS increment (knots)
$V_{cl,2}$ = standard climb CAS between 10,000 ft and Mach transition altitude

Since all of the speed parameters are provided as CAS, they are converted to true air speed (TAS) for use in SAGE using the conversion equations provided in the BADA User Manual [EEC 2004]. With no winds currently modeled (i.e., wind speed = 0), true air speed is effectively equivalent to ground speed in SAGE. However, per INM (SAE AIR 1845) specifications, a nominal 8 knot headwind is applied to all flights for altitudes up to 1000 ft. Although INM uses 8 knots throughout the LTO profiles, an engineering judgment was made to discontinue the use of this headwind in SAGE above 1000 ft. This is meant to be more conservative than continuing to use the 8 knot headwind throughout the LTO profiles at all airports.

The BADA User Manual also provides equations to calculate the transition altitude above which cruise speeds are used. In using Equation 12, the stall speed, $(V_{stall})_{TO}$, must be corrected for the mass of the aircraft as shown in Equation 14.

$$(V_{stall})_{TO,c} = (V_{stall})_{TO}(m / m_{ref})^{1/2} \qquad (14)$$

where $(V_{stall})_{TO,c}$ = corrected takeoff stall CAS (knots)
m = mass of current state of the aircraft
m_{ref} = reference mass of the aircraft (kg)

In order to make use of Equation 14, the starting mass of the aircraft must be calculated. The starting mass is the takeoff weight of the aircraft (from Section 3.3.1) minus any fuel burned during departure taxi and delay operations.

3.4.2.3 Calculate Lift and Drag

Lift and drag forces are computed using the data and equations prescribed by BADA [EEC 2004]. The lift and drag coefficient equations are reproduced from BADA as equations 15 and 16.

$$C_L = (2mg) / (\rho V_{TAS}^2 S \cos\Phi) \qquad (15)$$
$$C_D = C_{D0,CR} + C_{D2,CR} C_L^2 \qquad (16)$$

where C_L = lift coefficient
m = mass of aircraft (kg)
ρ = ambient air density (kg/m^3)
V_{TAS} = true air speed (m/s)
S = wing surface area (m^2)
Φ = flight path angle assumed to be zero (degrees)
C_D = drag coefficient for approach
$C_{D0,CR}$ = parasitic drag coefficient for cruise
$C_{D2,CR}$ = induced drag coefficient for cruise

Equation 16 is specific to the cruise mode and is presented as an example. Equations for the other modes are similar in form. When the induced and parasitic drag coefficients (e.g., $C_{D0,CR}$ and $C_{D2,CR}$) for modes other than cruise are not available, the current recommendation in the BADA user manual is to use the cruise coefficients. However, as discussed in Section 3.3.2, the scaling factors in Table 5 should be used to derive the missing coefficients to more accurately model drag.

A deficiency concerning the drag coefficient calculated from Equation 16 is that it does not account for compressibility effects (i.e., transonic drag rise effects). BADA currently does not include these effects. Therefore, the equation was modified to include a compressibility coefficient (ΔC_{DC}) as indicated in Equation 17.

$$C_D = C_{D0,CR} + C_{D2,CR} C_L^2 + \Delta C_{DC} \qquad (17)$$

The compressibility coefficient is based on an empirical fit originally developed by Stanford University which was modified to simplify the input requirements [Klima 2005]. Equations 18-24 describe the original fits used to calculate the compressibility coefficient.

$$X = M / M_{cc} \tag{18}$$
$$Y = X - 1 \tag{19}$$

If $X \geq 1.0$, then $\Delta C_{DC} = 0.001 + 0.02727Y - 0.1952Y^2 + 19.09Y^3$ (20)

If $1.0 > X \geq 0.95$ then $\Delta C_{DC} = 0.001 + 0.02727Y + 0.492Y^2 + 3.573Y^3$ (21)

If $0.95 > X \geq 0.8$ then $\Delta C_{DC} = 0.0007093 + 0.006733Y + 0.01956Y^2 + 0.01185Y^3$ (22)

If $0.8 > X \geq 0.5$ then $\Delta C_{DC} = 0.00013889 + 0.00055556Y + 0.00055556Y^2$ (23)

If $0.5 > X$ then $\Delta C_{DC} = 0$ (24)

Where ΔC_{DC} = compressibility coefficient
where X, Y = intermediate variables
M = Mach number
M_{cc} = crest critical Mach number

The crest critical Mach number (M_{cc}) is itself based on a complex empirical equation using difficult-to-obtain data: quarter-chord wing sweep, mean thickness to chord ratio, and the supercritical factor. An analysis showed that M_{cc} could be approximated by replacing this difficult-to-determine variable (due to the data requirements) with the BADA aircraft-specific average cruise Mach number. This simplification allowed the incorporation of compressibility effects in SAGE.

Once the lift and drag coefficients are calculated from equations 15 and 17, the actual lift and drag forces are calculated from equations 25 and 26 [EEC 2004].

$$L = (C_L \rho V_{TAS}^2 S) / 2 \tag{25}$$
$$D = (C_D \rho V_{TAS}^2 S) / 2 \tag{26}$$

where L = lift force (N)
D = drag force (N)

3.4.2.4 Calculate Thrust and Rate of Climb

As mentioned in Section 2.3, the underlying performance model involves balancing all the forces acting on an aircraft through the TEM as indicated by Equation 1 which is reproduced below:

$$(T - D)V_{TAS} = mg\frac{dh}{dt} + mV_{TAS}\frac{dV_{TAS}}{dt} \tag{1}$$

In addition to all the variables that have previously been defined, the dh/dt term is equivalent to the ROC. As a result, there are essentially three independent variables represented in this equation: speed, thrust, and ROC. In general, the most common use of this equation involves obtaining speed and thrust from other sources and calculating ROC from the equation. This is the overall method employed within SAGE whenever the ROC cannot be determined independently.

For takeoff/climbout, Equation 4 from INM is rearranged to calculate the uncorrected thrust based on the use of regression coefficients corresponding to maximum power during takeoff. The resulting equation is presented as Equation 27.

$$F_{max} = (E + F v + G_A h + G_B h^2 + H T_C)\delta \tag{27}$$

where F_{max} = uncorrected thrust corresponding to maximum power during takeoff for

one engine
δ = pressure ratio (altitude to sea-level)
v = equivalent/calibrated airspeed (knots)
h = pressure altitude Mean Seal Level (MSL) (feet)
T_C = temperature at the aircraft (°C)
E, F, G_A, G_B, H = regression coefficients that depend on power state (max-takeoff or max-climb power) and temperature state (below or above engine breakpoint temperature)

For clarity, the uncorrected thrust corresponds to at-altitude conditions whereas the corrected thrust in Equation 4 corresponds to the ISA at sea-level conditions. In SAGE, maximum power is assumed for the takeoff mode up to 2000 ft at which point, reduced power is modeled. The thrust for climbout is modeled essentially using the same equation (Equation 27) but with coefficients corresponding to the climbout mode.

Once the thrust has been determined, the ROC is calculated from the energy equation (Equation 1). Equation 1 is simplified through the definition of an Energy Share Factor (ESF) as shown in Equation 28.

$$ESF = [1 + (V_{TAS}/g)(V_{TAS}/dh)]^{-1} \qquad (28)$$

where   ESF = energy share factor

As prescribed in the BADA User Manual [EEC 2004], the following values for ESF are used:

- Acceleration in climb: ESF = 0.3
- Deceleration in descent: ESF = 0.3
- Deceleration in climb: ESF = 1.7
- Acceleration in descent: ESF = 1.7

In addition to the ESF, an adjustment is also made to model reduced climb power as indicated by Equation 29.

$$C_{pow,red} = 1 - C_{red}[(m_{max} - m_{act}) / (m_{max} - m_{min})] \qquad (29)$$

where $C_{pow,red}$ = reduced climb power coefficient
C_{red} = maximum reduction in power coefficient
m_{max} = maximum aircraft mass (kg)
m_{min} = minimum aircraft mass (kg)
m_{act} = actual (or current) aircraft mass (kg)

As prescribed by the BADA User Manual [EEC 2004], the C_{red} values are based on aircraft cateories:

- Turboprops: C_{red} = 0.25
- Pistons: C_{red} = 0
- Jets: C_{red} = 0.15

Application of the ESF and the reduced climb power coefficient in the energy equation to calculate ROC results in Equation 30.

$$ROC = [(F - D)V_{TAS}C_{pow,red}ESF / mg] \qquad (30)$$

For approach (or descent), thrust is calculated from BADA equations of the type reproduced as Equation 31.

$$\text{If } h > h_{des}, \text{ then } T_{des,high} = C_{tdes,high} T_{max\ climb} \tag{31}$$

where $T_{des\ high}$ = Descent thrust
h_{des} = transition altitude for initial descent from BADA
$C_{tdes,high}$ = high altitude descent thrust coefficient
$T_{max\ climb} = F_{max}$

The coefficient ($C_{tdes,high}$) in Equation 28 will change depending on altitude due to configuration changes. Unlike the takeoff/climbout modes, the calculation of the ROC for the approach mode is based on a mirroring of the standard 3 and 5 degree glides slopes used for jets and turboprops, respectively, in INM. Equation 32 exemplifies the ROC calculation for jets.

$$ROC = -\sin(3°)V_{TAS} \tag{32}$$

Once the thrust and ROC values are calculated, horizontal distance equations from INM are used to verify that the changes in altitudes for the takeoff/climbout chords as defined by the placeholders indicated in Figure 5 are consistent with the values calculated through the INM distance equations which take into account the calculated thrust and ROC values. Specifically, the INM distance (s_i) equations shown in Appendix M are used to calculate the horizontal distance for each chord. This distance is then used to calculate a change in altitude from the previously determined ROC and speed values. This change in altitude is compared to the value calculated from the placeholder altitudes according to equations 33-35.

If $|\Delta h_{Placeholder} - \Delta h_{calc}| \leq 1$, then no changes are necessary (33)
If $\Delta h_{Placeholder} > \Delta h_{calc} + 1$, then $h_{tn} = h_t - (|\Delta h_{calc} - \Delta h_{Placeholder}| / 2)$ and recalculate ROC, thrust, etc. (34)
If $\Delta h_{Placeholder} < \Delta h_{calc} + 1$, then $h_{tn} = h_t + (|\Delta h_{calc} - \Delta h_{Placeholder}| / 2)$ and recalculate ROC, thrust, etc. (35)

where $\Delta h_{Placeholder}$ = change in altitude determined from placeholder altitudes (m)
Δh_{calc} = change in altitude calculated from ROC, thrust, etc. (m)
h_t = altitude of tail of chord(m)
h_{tn} = altitude of new tail of chord(m)

Equations 34 and 35 essentially represent algorithmic loops that force the altitudes to converge on agreeable values. Because the criteria used for the comparisons is 1 m and because convergence generally occurs very quickly, any deviations from the placeholder altitudes are generally very small.

For cruise, the ROC is calculated from the ETMS radar data or from the OAG assigned trajectories. Distances and changes in time are also calculated from the trajectory and speed data. For OAG flights, ROC is always zero during cruise since altitude is modeled as a constant. Once the ROC is known, Thrust is calculated using Equation 1. If ROC is zero, then thrust is set equal to the drag force.

3.4.3 Fuel Flow and Fuel Burn Computations

Fuel flows are calculated using the BADA thrust specific fuel consumption (TSFC) equations [EEC 2004]. Rearranging these equations to directly include thrust, the fuel flow models for jet aircraft are presented as equations 36 and 37.

Takeoff and Approach: $\quad f = C_{f1}(1 + (V_{TAS} / C_{f2})]T$ (36)
Cruise: $\quad f = C_{f1}(1 + (V_{TAS} / C_{f2})]TC_{fcr}$ (37)

where f = fuel flow (kg/min)
C_{f1} = first thrust specific fuel consumption coefficient (kg/min/kN)
C_{f2} = second thrust specific fuel consumption coefficient (knots)
C_{fcr} = cruise fuel flow correction coefficient
C_{f3} = first descent fuel flow coefficient (kg/min)
C_{f4} = second descent fuel flow coefficient (ft)

Equation 36 is essentially identical to Equation 2 and is reproduced here for clarity. As indicated, the fuel flow equations are dependent only on velocity and thrust. Although BADA prescribes the use of a low thrust fuel flow equation to model descent and idling modes, Equation 36 is used to model approach since it seems to produce more reasonable results from comparisons with the ICAO databank fuel flow values.

Because power is assumed to remain constant at 7% thrust during taxi and delay (idling) operations, the standard fuel flows from the ICAO emissions databank for that power level were used instead of a BADA fuel flow equation. These ICAO fuel flows are adjusted for temperature, pressure, and Mach number according to the BFFM2 prescribed equation (see Section 3.4.4).

Once the fuel flows are calculated, fuel burn is determined by multiplying fuel flow by the change in time for the flight chord (or the duration associated with a taxi/delay period). The resulting finite amount of fuel burned for the flight chord (or taxi/delay period) is used to debit the aircraft weight before moving onto the next chord. Therefore, the aircraft experiences a reduction in weight as it moves along its flight path.

The amount of fuel burned for each flight is multiplied by appropriate unscheduled and canceled flights scaling factor (see Section 3.3.4). Although this can be accomplished through post-processing steps (e.g., for flight-level inventory development), this is currently accomplished directly within the FBE module to reduce the number of computational steps. The scaling is also performed for the emissions calculated in Section 3.4.4. If necessary, the effects of these scale factors can be readily removed through post-processing efforts.

3.4.4 Emissions Modeling

SAGE generates emissions of CO, HC, NOx, CO_2, H_2O, and SOx (modeled as SO_2) for each flight chord. CO, HC, and NOx emissions are modeled through the use of BFFM2 [Baughcum[a] 1996]. CO_2, H_2O, and SOx emissions are modeled based on fuel composition using Boeing-derived emissions indices [Hadaller 1989 and 1993]. These methods and data have been used in various past inventory projects, examples of which can be found in Baughcum[a,b] 1996, Baughcum[c,d] 1998, Daggett 1999, Sutkus Jr. 2001, and IPCC 1999. A review of BFFM2 by the Alternative Emissions Methodology Task Group (AEMTG) of ICAO CAEP Working Group 3 (WG3) resulted in the conclusion that the method is an acceptable approach for modeling emissions of CO, HC, and NOx [ICAO[b] 2004]. Partly because of this type of recognition and also because it is an open and publicly available method, BFFM2 has seen increased international use in recent years. Therefore, BFFM2 can be considered the de facto standard in modeling aircraft emissions.

3.4.4.1 Modeling CO, HC, and NOx Emissions

BFFM2 involves the use of the emissions indices (EI) and corresponding fuel flows from the ICAO databank to model CO, HC, and NOx emissions [ICAO 1995 and ICAO[a] 2004]. The key components within BFFM2 are the adjustments for altitude (e.g., atmospheric effects) and the development of a relationship between EI and fuel flow. The latter component allows a predicted fuel flow from BADA to be used to predict emissions. For clarity, the predicted fuel flow and emissions corresponding to at-altitude conditions will hereinafter be referred to as "uncorrected" values while the fuel flow and emissions in the ICAO databank will be referred to as "corrected" because they correspond to the reference condition: international standard atmosphere (ISA) at sea-level. Several distinct steps are required to model emissions.

In the first step, the four ICAO corrected fuel flows are adjusted for installation effects. That is, each of these fuel flows is multiplied by an adjustment (or correction) factor. The general factors provided by Boeing are used in the absence of fuel flow values corresponding to an engine-installed-on-an-aircraft situation [Baughcum 1996]. These factors are used according to Equation 38.

$$RW_{ff} = RW_{ffu} r \qquad (38)$$

where RW_{ff} = corrected and adjusted (for installation effects) fuel flow corresponding to the reference condition (kg/s)
RW_{ffu} = corrected but unadjusted (for installation effects) fuel flow from ICAO databank corresponding to the reference condition (kg/s)
r = Boeing adjustment/correction factor
= 1.010 (takeoff)
= 1.013 (climbout)
= 1.020 (approach)
= 1.100 (idle)

For the second step, log-log plots between the ICAO corrected EIs and fuel flows are developed. As exemplified in Figure 7, the plot for NOx is developed using a regressed linear fit whereas the CO and HC fits are based on a bilinear approach.

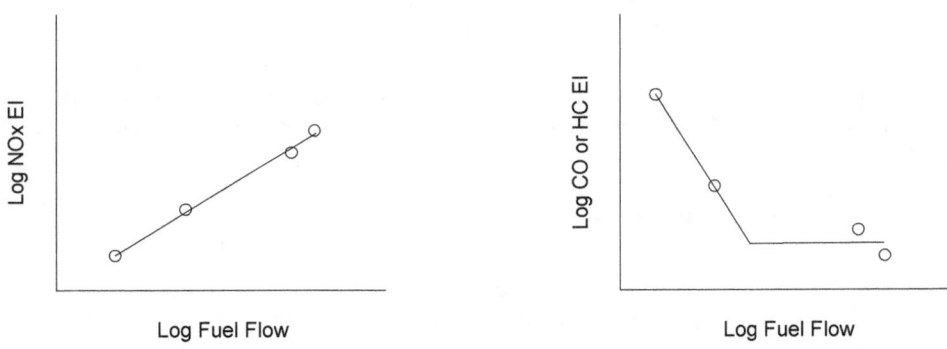

Figure 7. Log-Log Plots of EI versus Fuel Flow for NOx, CO, and HC

For this bilinear approach, a line is created between the two lower power setting points. The second line is horizontal and positioned in the middle of the two vertical positions of the higher power settings. Then the two lines are simply extended toward each other until they intersect. All of these plots are modeled programmatically using the log values and fit-parameters specified in Section 3.3.3.

In the third step, the uncorrected BADA fuel flow is corrected to the reference condition using the prescribed BFFM2 equation [Baughcum 1996] reproduced as Equation 39.

$$W_{ff} = (W_f / \delta_{amb})[\theta_{amb}^{3.8} \exp(0.2M^2)] \qquad (39)$$
$$\delta_{amb} = P_{amb} / 14.696$$
$$\theta_{amb} = (T_{amb} + 273.15) / 273.15$$

where W_{ff} = corrected BADA fuel flow (kg/s)
W_f = uncorrected BADA fuel flow (kg/s)
P_{amb} = at-altitude ambient pressure (psia)
T_{amb} = at-altitude ambient temperature (°C)
M = Mach number

The fourth step involves the use of the corrected fuel flow and the log-log plots to determine the corresponding EI values. Again, this is conducted programmatically using the linear log-log fitted equations rather than actual plots. For the fifth step, the calculated EI values are uncorrected using the BFFM2 prescribed equations [Baughcum 1996] in order to reflect the at-altitude flight conditions:

$$EIHC = REIHC(\theta_{amb}^{3.3} / \delta_{amb}^{1.02}) \qquad (40)$$
$$EICO = REIHC(\theta_{amb}^{3.3} / \delta_{amb}^{1.02}) \qquad (41)$$
$$EINOx = RENOx[\exp(H)](\delta_{amb}^{1.02} / \theta_{amb}^{3.3})^{1/2} \qquad (42)$$
$$H = -19.0(\omega - 0.0063) \qquad (43)$$

where EIHC = uncorrected emissions index for HC (g/kg)
EICO = uncorrected emissions index for CO (g/kg)
EINOx = uncorrected emissions index for NOx (g/kg)
REIHC = corrected emissions index for HC (g/kg)
REICO = corrected emissions index for CO (g/kg)
REINOx = corrected emissions index for CO (g/kg)
H = humidity correction factor
ω = specific humidity

The complete set of equations used to calculate specific humidity (ω) is not shown for simplicity. In the sixth step, the time duration for the flight chord is determined and then used to calculate fuel burn as shown in Equation 44.

$$FB = f_{ua}\Delta t \qquad (44)$$

where FB = fuel burn (kg)
f_{ua} = uncorrected and adjusted (for installation effects) fuel flow (kg/s)
Δt = time duration of flight chord (s)

Lastly, the fuel burn and uncorrected EIs are used to calculate emissions for CO, HC, and NOx as indicated by Equation 45 using NOx as an example.

$$EN = (FB)(EINOx) \qquad (45)$$

where EN = emissions of NOx (g)

The published writeup of the BFFM2 methodology [Baughcum] provides an overall description of the general situation when the engine data in the ICAO databank behaves according to the prescribed methodology. However, there are several instances where the data behaves differently than exemplified in Figure 7. The BFFM2 writeup appears to recognize that these cases exist, but no specific guidance is provided on how to handle them. These issues have also been documented within an ICAO CAEP WG3 paper [ICAO 2005]. The issues and the current solutions are described in Table 7.

Table 7. Issues Associated with the use of BFFM2 and the ICAO Emissions Databank

Issue	Description	Solution Currently Employed in SAGE Version 1.5
1	For some engines, the slanting line of the bilinear fit for CO and HC intersects the horizontal line beyond the 85% power point.	Extend the slanting line from the 30% point until the fuel flow value of the 85% point is reached. Then move vertical downwards until the horizontal line is intersected. This results in a vertical intermediate line between the two "normal" lines.
2	For some engines, the horizontal line of the bilinear fit for CO and HC intersects with the slanting line above the 30% point.	Create a horizontal line at the 30% point and use that line to form the bilinear fit.
3	For some engines (e.g., those with dual annular combustors), the slanting line of the bilinear fit for CO and HC has an increasing slope.	Use the horizontal line formed by the higher power points to conduct all of the modeling. This was done for simplicity due to the fact that the two lines do not even intersect within the normal LTO power range (i.e., 7% to 100%).
4	In a few cases, the EI and/or fuel flow values may not be available for some engines in the ICAO databank.	The missing data points were assigned zero values and plots developed accordingly.
5	Due to the asymptotic nature of the bilinear fit for CO and HC, low fuel flows can result in unreasonably high EI values.	A lower limit on fuel flow was placed corresponding to the fuel flow from the ICAO databank for the 7% (idle) power condition. This was done globally for engines.
6	In the ICAO databank, some EI values are zeros which cannot be modeled on log plots.	The zero values were replaced by reasonably low values of 10^{-6}.
7	For some engines, the EI values are all zeros. As a result, HC and CO plots cannot be developed.	Emissions are modeled as zeros.

A list of affected engines is provided in Appendix N. This list shows all of the issues affecting each of the engines listed. Since Issue 5 potentially affects all engines, it is not included in Appendix N. Although a lot of engines appear to be affected, they generally account for a minority of total flights worldwide, and hence, a smaller part of the overall emissions. Notwithstanding these issues, BFFM2 still provides a solid foundation for modeling emissions. Hence, the solutions presented in Table 7 can be thought of as supplementary details necessary to apply BFFM2 to all engines globally. These solutions can be further improved through a better understanding of the ICAO data and engine behavior. To that end, some of the recommendations provided in ICAO 2005 are improvements over the current solutions which will be changed accordingly once an updated and standardized writeup of BFFM2 becomes available.

3.4.4.2 Modeling CO_2, H_2O, and SOx Emissions

Emissions of CO_2, H_2O, and SOx are modeled based on jet fuel composition. Boeing conducted a review of the available fuel composition data and developed emissions indices that could be used to predict emissions solely based on fuel burn [Hadaller 1989]:

- $EICO_2$ = 3155 g/kg
- EIH_2O = 1237 g/kg
- $EISOx$ (as SO_2) = 0.8 g/kg

The underlying assumption with this approach is that the fuel experiences 100% combustion. That is, all of the carbon (C) and hydrogen (H) within the fuel are converted to CO_2 and H_2O as exemplified in Equation 46.

$$C_mH_n \text{ (fuel)} + (m + n/4)O_2 \longrightarrow mCO_2 + (n/2)H_2O \qquad (46)$$

The complete combustion assumption is valid since modern jet engines are very efficient in maximizing fuel use. Furthermore, due to the fact that the mass of the emitted CO_2 and H_2O are significantly greater than that for CO, HC, or NOx, it can be assumed that all of the C and H are converted to CO_2 and H_2O (i.e., none left over for CO and HC). The error associated with making this assumption is likely very small (probably much less than 1%).

Similar to CO_2 and H_2O, the EI for SOx was derived in a similar way assuming that all of the sulfur (S) within the fuel is converted to SO_2. For simplicity, this is not shown in Equation 46. Indeed, sulfur is not the only other constituent within jet fuel; nitrogen (N) and trace amounts of metals can be found within jet fuel (although the N in NOx mostly comes from N_2 in the air). Although the derivation of EISOx is based on converting sulfur in the fuel to SO_2, it is approximated as SOx since all of the sulfur has been accounted for. Due to the small content of sulfur within jet fuel, modeling SOx emissions is not likely to be as accurate as modeling CO_2 or H_2O emissions.

3.4.5 Additional Data Checks

In addition to the data processing described in Section 3.1, a couple of data checks have been implemented directly within the FBE module. The first involves a check of each chord's distance to catch bad ETMS data through the use of the following criteria: A single chord should not have a distance greater than the Great Circle distance of the corresponding flight. Although the various higher-level checks conducted in Section 3.1 would likely have captured all or most of the bad ETMS distance data, this explicit check of each chord serves as an extra quality check. Although it is theoretically possible for a single chord to have a distance greater than the Great Circle distance of the entire flight, it is highly unlikely since it would require unusual circumstances such as severe detours or re-routes. Only a small fraction (much less than 1%) of ETMS flights actually triggers this criteria.

The second data check involves ETMS speed data. An algorithm was implemented to smooth out any erratic ETMS speed data. Except for the first and last chords, each chord's Mach value speed is checked to see if the difference between the current and previous chords is less than 0.2 or if the current chord is less than 0.65 times or greater than 1.1 times the average aircraft-specific BADA Mach number. If any of these conditions are met, then the following chords' speeds are checked one at a time until a suitable speed is found, in which case, the suitable speed is assigned to the current chord. This check actually helps to correct (smooth out) erroneous ETMS speed data.

4 METHODOLOGY – FORECASTING

As indicated in Figure 8, the forecasting module within SAGE requires, among others, the flight forecasts from the FAA's TAF and the FESG's projections.

Figure 8. Overview of Forecasting within SAGE

Figure 8 is actually an exapansion of the simplified forecasting component depicted in Figure 3. The overall method involves growth of a week's worth of a base year's OAG scheduled flights to represent the growth in demand for a future year. A week is used for consistency with the capacity and delay modeling described in Section 3.2.5, and again, the week is May 29 to June 4 based on the statistical analysis presented in Appendix J. Also included in the forecasting methodology is an aircraft retirements and replacements algorithm that uses retirement parameters from FESG. The result is a future schedule of flights that reflect the effects of growth and retirements/replacements. Additional details concerning the forecasting methodology are provided in the form of pseudo-code in Appendix O.

Both the week's worth of baseline and future schedules are run through the fuel burn and emissions module similar to the development of historical inventories. The results from the baseline and future schedules are compared to determine relative statistics such as percent differences. These weeks' worth of results are never scaled to determine magnitudes of fuel burn and emissions. Rather, the percent differences are applied to the historical inventory(ies) of the baseline year in order to obtain more accurate results. This approach of only using a week's worth results for comparison purposes minimizes discrepancies associated with using scheduled flights. Also, applying the percent differences to the historical baseline makes the forecasted magnitudes consistent with the historical baseline. The latest baseline year used in SAGE is 2002 because the most recent FESG data also uses that year as a baseline.

The forecasting methodology discussed in this section reflects the development of a "standard" SAGE forecast of fuel burn and emissions. Deviations from this scenario (i.e., non-standard scenarios) can be modeled through modifications of various parameters including those related to aircraft performance, aircraft/engine substitutions, schedules, trajectories, growth factors, retirement parameters, etc. These changes are possible because the parameters are available through input data tables rather than being hard-coded within the model. Similar to the FBE module, this capability provides the user with a dynamic modeling environment in which various what-if scenarios can be analyzed. Although the burden is still on the user to make these changes intelligently, the availability of the parameters allow for a high-degree of flexibility and control in modeling various scenarios.

The following sections provide details of the forecasting process: (1) growth of scheduled flights (2) retirements and replacements, and (3) application of baseline and forecasted schedules.

4.1 Growth of Scheduled Flights

Growing flights to represent future demand is based on the use of FAA's TAF and the FESG's projected number of future flights. The TAF data is considered to be more accurate than the FESG data in large part because TAF projections are airport based whereas FESG projections are based on large-scale regional and route categories. Therefore, TAF data takes precedence over the FESG data when modeling US flights covered under the TAF. Flights for the remaining US flights and flights in the rest of the world are based on FESG data. Growing a flight schedule requires the following steps: (1) TAF to grow US flights, (2) FESG to grow non-US flights, and (3) modeling in WWLMINET.

4.1.1 TAF to Grow US Flights

The use of airport-based TAF operations involves the use of a Fratar algorithm [Long 1999] which is the most widely used method for generating trips based on the TAF. Both the DOT and FAA have used it in their transportation planning models. The Fratar algorithm is a trip distribution method that scales up the baseline matrix of OD operations to meet the TAF's future operations. That is, the future airport operations are used to create a future schedule of flights based on the distributions specified by the baseline year schedule. The result from the use of the TAF and the Fratar algorithm is a future schedule with multipliers for each flight to indicate the growth in flights.

4.1.2 FESG to Grow Non-US Flights

Since the flights covered by the TAF are generally only US to US flights, the FESG data is used to grow the remaining flights. This is accomplished through the use of growth factors from the FESG forecasted number of flights. For each route group identified in the FESG data, a growth factor is developed by dividing the FESG forecasted number of flights by the FESG baseline year's number of flights. This factor is then applied to each scheduled baseline flight that falls within the corresponding route group. Similar to the TAF-Fratar method, the result is a future schedule for the remaining flights with multipliers.

4.1.3 Modeling in WWLMINET

These non-integer multipliers cannot be used in SAGE because WWLMINET requires individual whole "ticket" listings of flights (i.e., fractional flights cannot be used by WWLMINET's queuing algorithm). Therefore, fractional contributions of similar flights are summed to obtain whole numbers and rounded to prevent any leftovers.

The resulting future schedule and the baseline schedule are modeled in WWLMINET to obtain the various terminal area delays (see Section 3.2.5). The delays for an airport in the future year should always be equal to or greater than the delays for the same airport in the baseline year. This is assuming that the airport has a greater number of flights in the future year. The queue modeling in WWLMINET guarantees that depending on the growth, the delays will increase accordingly.

In addition to these delays, WWLMINET also generates indicators that specify which flights need to be canceled due to excessively long delays. Currently, the delay criteria is set at 2 hours; any delay greater than 2 hours results in a cancellation of the flight. The canceled flights are removed from the forecasted schedule. Then the individual whole number flights are aggregated such that identical flights are represented as a single record with a flight multiplier.

4.2 Aircraft Retirements and Replacements

The population-based retirement parameters cannot be applied directly to the forecasted schedule because the age of the aircraft listed in the schedule is unknown. Therefore, the BACK world fleet database is used to develop airline and aircraft specific "conversion" factors that encompass the effects of these retirement parameters. This method is based on the use of individual aircraft listings in the BACK fleet rather than in groups since it allows finer control over the various aircraft properties such as age, airline, country affiliation, number of seats, etc. The most important control is the ability to age individual aircraft in yearly increments including new aircraft that come into service. To implement this approach, several distinct steps are required: (1) set up baseline fleet, (2) create future aircraft, (3) creation of forecasted fleet, (4) application of retirement parameters, (5) development of conversion factors, and (6) application of conversion factors.

4.2.1 Create Baseline Fleet

The baseline fleet is set up by using the specific BACK world fleet data of "active" aircraft that correspond to the baseline schedule year. Individual aircraft records (or objects) are created by extracting aircraft code, aircraft age and various categorization data such as airline, country affiliation, number of seats, etc. from the BACK data. For the airline category, the top 50 and a generic 51st category are created which includes all others. The extracted data is used to further categorize and clean the data such as matching aircraft codes to those within the OAG schedule and assigning seat categories (i.e., bins). Also, a count property is assigned to each aircraft with a value of one. Further details can be found in Appendix O.

4.2.2 Create Future Aircraft

Distributions of replacement aircraft types are developed by processing just the BACK aircraft that are inactive with one of the following status labels:

- On Order
- On Option
- Pre-Service Built

Aircraft within these categories represent those that have or will be purchased by an airline for entry into service in the "near" future (e.g., "guaranteed orders"). As a result, they represent the current state-of-the-art aircraft. These "future" aircraft are categorized by aircraft code and seat categories, and then counted in order to develop distributions which are used to replace retired aircraft and also to assign new aircraft for fleet growth.

4.2.3 Creation of Forecasted Fleet

An incremental future fleet experiencing only growth is developed from the baseline fleet. A copy of the baseline fleet is made and each aircraft is aged by one year. Incremental growth factors are developed from the FESG forecasted and baseline fleet counts. The growth factors are applied to each individual aircraft in the future fleet based on seat category. The result is a set of new aircraft records with the same properties as the existing aircraft but with different aircraft types and fractional values assigned to their count properties. Each new aircraft are also assigned an age of 0.5 yr as an approximation to indicate that it entered into service in the middle of the year. Aircraft types (aircraft codes) are assigned to these new aircraft based on randomly selecting from the distribution of future aircraft types by seat category. Depending on the country affiliation of each new aircraft, either all or part of the new aircraft is based on the future aircraft type. If a new aircraft's affiliated country has adopted the noise chapter 2 phase-out rule, it will experience the use of 100% future aircraft types. In contrast, if the new aircraft's affiliated country has not adopted the phase-out rule, only a portion of the new aircraft will based on a future aircraft type. The remaining portion will be based on the same aircraft type as the existing aircraft from which it was copied form. This means that two new aircraft will actually be created since one is based on a future aircraft type and the other on the existing type. This split into two aircraft records is accomplished through the use of the year-specific phase-out fractions which are reflected accordingly by each aircraft's count property. As indicated in Section 2.13, application of the phase-out rule and fractions mimics their use in MAGENTA. This third step is repeated iteratively on a yearly basis aging each aircraft accordingly until the target future year is reached. The result is a large population (a grown fleet) of individual aircraft records with varying count values.

4.2.4 Application of Retirement Parameters

The retirement parameters are applied to a copy of the grown fleet by first identifying the aircraft category in Section 2.14 that each of the aircraft records falls into, and then using the corresponding retirement parameters from Table 2 to apply the survivor curves represented generically as equation 47 to each aircraft record.

$$Sr = constant + ax + bx^2 + cx^3 + dx^4 + ex^5 + fx^6 \qquad (47)$$

where Sr = survivor factor (fraction of aircraft/fleet that survived the retirement process)
x = age of aircraft
constant, a, b, c, d, e, f = retirement parameters specified in Table 2.

Since Equation 47 represents the survivor fraction, 1-Sr represents the retired and replaced fraction. Therefore, each aircraft affected by these curves (i.e., within the range of applicable years shown in Table 2), experience a reduction in its count property by 1-Sr. This reduction is exactly counter-balanced by the addition of a new aircraft record. This new aircraft is treated identically to the one created through growth effects discussed in the third step including the use of the phase out rule and fractions. The result is a fleet that has undergone growth and retirements.

4.2.5 Development of Conversion Factors

Conversion factors are developed that represent a ratio of the effects of fleet growth and retirements to growth only. To accomplish this, the count property of each aircraft record within the resulting fleets from the third and fourth steps are summed based on a combination of aircraft type and airline categories. These counts are used to develop the conversion factors as shown in Equation 48

$$CF = nGR / nG \qquad (48)$$

where CF = conversion factor
nG = number of aircraft in each category due to growth only
nGR = number of aircraft in each category due to growth and retirement

Due to the categorizations used for the counts, the conversion factors are based on aircraft type and airline categories.

4.2.6 Application of Conversion Factors

The conversion factors are applied to the forecasted schedule by multiplying each flight multiplier from Section 4.1.3 by the corresponding conversion factor based on aircraft type and airline. Thus, new flight multipliers are generated. As the name implies, the conversion factors convert the grown-only flights to include the effects of retirements (both growth and retirements).

Although a disconnect exists between the growth of the fleet and the growth of the schedules, the application of the conversion factors to the schedules is justified because the growth of the schedules (from the FESG flight projections) took into account aircraft seat size and range. Therefore, the relative strengths of the conversion factors by aircraft type will generally correspond to the relative strengths of the flight multipliers by aircraft type. That is, the aircraft with the higher conversion factors will be reflected by a similar prominence of flights (as reflected by the flight multipliers) for the same or similar aircraft types.

It is possible that an aircraft type and airline category combination from the fleet doesn't exist in the schedule due to the fact that some on-order aircraft types have no flights listed in the baseline year schedule. If this is the case, then additional flights need to be created in order to preserve the aircraft counts created during the aging and retiring process.

The solution employed in this case is to develop a ratio of counts for the aircraft type in question with the most popular (highest count) aircraft within the same airline and range categories. If the airline category cannot be matched, then the 51st category should be used. This ratio should be less than 1 unless the aircraft in question is the most popular, in which case the ratio will be based on a comparison with the next most popular. Additional flights are created in the schedule by copying flights with the most popular aircraft (or the next most popular) for that airline. The aircraft types are replaced by the new aircraft and the flight multipliers are multiplied by the aforementioned ratio in order to reflect the flights of the new aircraft.

4.3 Application of Baseline and Forecasted Schedules

Once the baseline and the final forecasted schedule have been developed, they are run through the FBE module similar to the movements database in developing a historical inventory. Essentially, the only difference is the additional use of the multiplying factors which have the effect of scaling the results. Once the results are obtained, there are two general ways to use them: (1) comparative analyses and (2) development of future magnitudes.

4.3.1 Comparative analyses

The fuel burn and emissions totals from the baseline and forecasted schedules are used to calculate comparative metrics such as percent differences and ratios. Both are typically used to quantify the relative effects of the forecasted scenario. They show the percent or fractional increase in loading of fuel burn and emissions for the future year. Although the standard forecasts generated by SAGE will result in increased loadings, non-standard modeled scenarios involving changes to aircraft parameters, growth rates, schedules, etc. can result in decreases as well.

The comparative metrics are most accurate when they are based on global totals. Regional, country, and even airport-level differences can be analyzed, but the generalized nature of the input data (TAF, FESG, etc.) imply that caution must be exercised when reviewing metrics at these levels. The data as well as model assumptions need to be carefully considered to justify finer level comparisons. In general, global and regional levels are considered large enough to conduct assessments without any real concern.

4.3.2 Development of Future Magnitudes

To model future magnitudes, the ratios of the future to baseline results are applied to the appropriate SAGE historical inventories (e.g., 2002 inventories). As discussed in Section 4.3.1, global and regional results can generally be used safely due to their scales. Therefore, the global and regional ratios are multiplied by the global and regional inventories to obtain the corresponding future magnitudes of fuel burn and emissions. This approach essentially has the effect of using the growth factors from TAF and FESG but the baseline movements data from SAGE. The assumption here is that the inclusion of ETMS data and unscheduled flight scaling factors, SAGE movements data represents a more accurate accounting of baseline flights than either the TAF or the FESG data, although they should all comfortably agree.

5 OUTPUT RESULTS

As shown in Figure 3, SAGE generates three raw inventories and then uses them to generate several processed (queried) inventories. These inventories are discussed in the following sections.

5.1 SAGE Raw Inventories

The three inventories that SAGE generates are: (1) flight-level model, (2) chord-level, and (3) 4D world grids.

5.1.1 Raw Flight-Level Modal Inventory

The flight-level modal inventory contains listings of each individual flight on a global basis. This inventory contains over 30 millions per year. Essentially, all the flights within the movements database are reproduced but with fuel burn and emissions results added. These results are provided modally as indicated by the fields shown below:

- flight_key = unique SAGE flight key
- flight_date = flight departure date
- source_flag = E=ETMS, O=OAG
- flight_id = flight ID
- status_flag = internal debugging flag
- dep_airport = departure airport code
- arr_airport = arrival airport code
- dep_time = departure time
- arr_time = arrival time
- cruise_altitude (ft) =cruise altitude
- track_no = dispersion track number for OAG flights
- aircraft_type = aircraft code
- aircraft_category = J=Jet, T=Turboprop, P=Piston
- num_engines =number of engines
- back_engine = BACK engine name/code
- icao_edms_engine = ICAO/EDMS engine name/code
- gc_distance (nm) = Great Circle distance
- flight_distance (nm) = flight distance
- carrier_code = carrier code
- carrier_name = carrier name
- region_code = region code (8 world regions)
- region_name = region name (8 world regions)
- region_end_type = I=flight ended in same region, O=ended elsewhere
- dep_country = departure country name
- arr_country = arrival country name
- takeoff_weight (kg) = assigned takeoff weight
- scale_factor = scale factor for unscheduled flights

- dep_gnd_distance (nm) = departure ground distance
- dep_gnd_fuelburn (kg) = departure ground fuel burn
- dep_gnd_co2 (g) = departure ground CO2
- dep_gnd_h2o (g) = departure ground H2O
- dep_gnd_sox (g) = departure ground Sox
- dep_gnd_co (g) = departure ground CO
- dep_gnd_hc (g) = departure ground HC
- dep_gnd_nox (g) = departure ground NOx
- to_co_distance (nm) = takeoff/climbout distance
- to_co_fuelburn (kg) = takeoff/climbout fuel burn
- to_co_co2 (g) = takeoff/climbout CO2
- to_co_h2o (g) = takeoff/climbout H2O
- to_co_sox (g) = takeoff/climbout Sox
- to_co_co (g) = takeoff/climbout CO
- to_co_hc (g) = takeoff/climbout HC
- to_co_nox (g) = takeoff/climbout NOx
- cruise_distance (nm) = cruise distance
- cruise_fuelburn (kg) = cruise fuelf burn
- cruise_co2 (g) = cruise CO2
- cruise_h2o (g) = cruise H2O
- cruise_sox (g) = cruise Sox
- cruise_co (g) = cruise CO
- cruise_hc (g) = cruise HC
- cruise_nox (g) = cruise NOx
- app_glide_distance (nm) = approach distance
- app_glide_fuelburn (kg) = approach fuel burn
- app_glide_co2 (g) = approach CO2
- app_glide_h2o (g) = approach H2O
- app_glide_sox (g) = approach Sox
- app_glide_co (g) = approach CO
- app_glide_hc (g) = approach HC
- app_glide_nox (g) = approach NOx
- arr_gnd_distance (nm) = arrival ground distance
- arr_gnd_fuelburn (kg) = arrival ground fuel burn
- arr_gnd_co2 (g) = arrival ground CO2
- arr_gnd_h2o (g) = arrival ground H2O
- arr_gnd_sox (g) = arrival ground SOx
- arr_gnd_co (g) = arrival ground CO
- arr_gnd_hc (g) = arrival ground HC
- arr_gnd_nox (g) = arrival ground NOx
- fuelburn (kg) = total fuel burn
- co2 (g) = total CO2
- h2o (g) = total H2O
- sox (g) = total SOx
- co (g) = total CO

- hc (g) = total HC
- nox (g) = total NOx

5.1.2 Raw Chord-Level Inventory

The chord-level inventory contains a listing of individual flight chord points for all flights worldwide resulting in approximately 1 billion yearly records. Even though each listing represents a point in space geometrically, it can be considered to represent a chord because much of the information provided in the inventory necessarily apply to the entire chord rather than just a point. For clarity, the ends of a chord are referred as either the head (beginning) or tail (ending) points. And it should be obvious that the tail point of one chord represents the head point of the next chord. Whether the data in the inventory applies to a point or the entire chord, the information is always stored at the tail point of the chord. The fields within this inventory are provided as follows:

- flight_key = unique SAGE flight key
- seq_no = chord sequence number
- mode = mode number for chord
- latitude (deg) = latitude of chord tail
- longitude (deg) = longitude of chort tail
- altitude (m) = altitude of chord tail
- chord_time = point in time at chord tail
- T_i (K) = temperature at chord tail
- P_i (Pa) = pressure at chord tail
- a_i (m/s) = speed of sound at chord tail
- M_i = average Mach number for chord
- h_i (m) = average height of chord
- delta_alt (m) = change in altitude for chord
- v_i (m/s) = average speed of chord
- delta_v (m/s) = change in speed for chord
- delta_t (s) = change in time for chord
- distance (nm) = length of chord
- thrust (N) = thrust for the chord
- weight (kg) = aircraft weight at chord tail
- CL_i = BADA lift coefficient for chord
- CD_i = BADA drag coefficient for chord
- L_i (N) = lift force for chord
- D_i (N) = drag force for chord
- f_i (kg/s) = fuel flow for chord
- percent_foo = percent power for chord
- REICO_i (g/kg-fuel) = corrected (reference) CO EI for chord
- REIHC_i (g/kg-fuel) = corrected (reference) HC EI for chord
- REINOx_i (g/kg-fuel) = corrected (reference) NOx EI for chord
- fuelburn (kg) = fuel burned for chord
- co2 (g) = CO2 emitted for chord
- h2o (g) = H2O emitted for chord
- sox (g) = SOx emitted for chord
- co (g) = CO emitted for chord
- hc (g) = HC emitted for chord

- nox (g) = NOx emitted for chord

5.1.3 Raw 4D World Gridded Inventory

The 4D world grid inventory contains a listing of flight segments similar to the chord-level inventory but the segments correspond to the portions of the chords that traversed a grid. The data is 4D since each segment listing contains flight date/time and grid location information. The similarity to the chord-level inventory is reflected by the approximately 900 million yearly records within this inventory. The fields in the inventory are shown below:

Flight_key = unique SAGE flight key
Flight_date = departure date
Track_id = dispersion track number for OAG flights
Seq_no = chord sequence number
Mode = mode number
i = latitude index
J = longitude index
k = altitude index
Time_in = time entered into grid
Deltat_i (s) = duration in grid
V_i (m/s) = average speed of chord
Fuelburn (kg) = fuel burned while in grid
co2 (g) = CO2 emitted while in grid
h2o (g) = H2O emitted while in grid
co (g) = CO emitted while in grid
sox (g) = SOx emitted while in grid
hc (g) = HC emitted while in grid
nox (g) = NOx emitted while in grid

The current standard grid size is 1° latitude by 1° longitude by 1 km altitude. But these specifications can be modified to obtain varying sizes in all three dimensions. The i, j, and k indices used in this inventory are based on the standard grid cell size, and their orientation is shown in Figure 8.

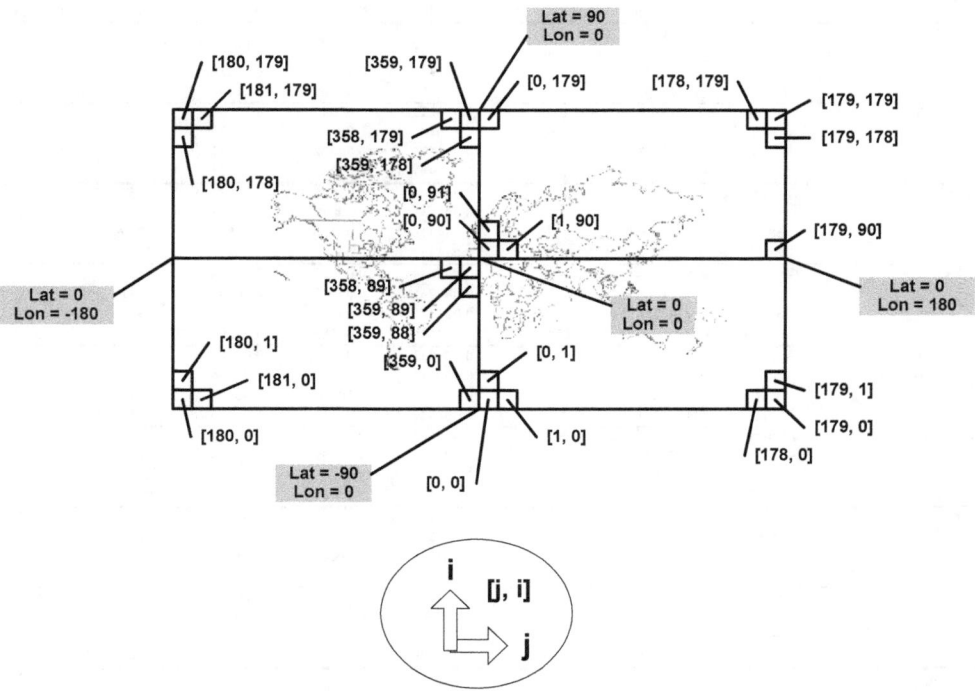

Figure 9. Orientation of Standard Grids within SAGE

5.2 Processed Inventories

Once the three raw inventories are available, they can be processed through various queries in order to obtain more useful inventories. As shown in Figure 3, some typical processed inventories include regional, country, aircraft-engine, and aggregated grid results. In each of these inventories, results from piston aircraft are currently excluded because the EI data for piston engines have been deemed unreliable. Since piston aircraft make up a very small fraction of total flights and distance traveled, this exclusion is reasonable. Figures 9-12 exemplify each of the processed inventories.

Region		Distance (nm)	Fuel Burn (Tg)	CO (Tg)	HC (Tg)	NOx (Tg)	CO2 (Tg)	H2O (Tg)	SOx (Tg)
Africa to Africa	Ground	2051164	0 04	0 000510	0 00010	0 00050	0 13	0 05	0 00003
	<=3000	6808762	0 08	0 000391	0 00007	0 00133	0 27	0 10	0 00007
	>3000	133618478	0 99	0 003385	0 00049	0 01218	3 13	1 23	0 00079
	Total	142478403	1 12	0 004285	0 00066	0 01401	3 53	1 38	0 00089
Africa to Other	Ground	465037	0 02	0 000225	0 00005	0 00035	0 06	0 02	0 00002
	<=3000	1628916	0 05	0 000187	0 00004	0 00116	0 17	0 07	0 00004
	>3000	159971053	2 34	0 004575	0 00072	0 03382	7 39	2 90	0 00187
	Total	162065007	2 42	0 004987	0 00080	0 03533	7 62	2 99	0 00193
Asia to Asia	Ground	15970660	0 53	0 005999	0 00091	0 00837	1 66	0 65	0 00042
	<=3000	51954044	1 23	0 004156	0 00048	0 02538	3 89	1 52	0 00099
	>3000	1336919880	15 01	0 034316	0 00410	0 22880	47 36	18 57	0 01201
	Total	1404844584	16 77	0 044471	0 00549	0 26255	52 91	20 74	0 01341
Asia to Other	Ground	1511632	0 09	0 001034	0 00025	0 00164	0 27	0 11	0 00007
	<=3000	4941110	0 26	0 000765	0 00013	0 00599	0 81	0 32	0 00021
	>3000	804202324	15 57	0 021727	0 00347	0 23530	49 11	19 25	0 01245
	Total	810655066	15 91	0 023526	0 00385	0 24293	50 19	19 68	0 01273
Australia and Oceania to Australia and Oceania	Ground	4744979	0 08	0 000937	0 00015	0 00102	0 24	0 09	0 00006
	<=3000	14162096	0 16	0 000611	0 00009	0 00286	0 51	0 20	0 00013
	>3000	315845206	2 15	0 004775	0 00088	0 02983	6 80	2 66	0 00172
	Total	334752281	2 39	0 006322	0 00112	0 03370	7 55	2 96	0 00191
Australia and Oceania to Other	Ground	370207	0 02	0 000203	0 00004	0 00039	0 06	0 02	0 00002
	<=3000	1213395	0 06	0 000151	0 00002	0 00134	0 18	0 07	0 00004
	>3000	163738686	2 78	0 003519	0 00077	0 04381	8 77	3 44	0 00222
	Total	165322218	2 86	0 003874	0 00084	0 04554	9 01	3 53	0 00228
Eastern Europe to Eastern Europe	Ground	2171666	0 05	0 001237	0 00043	0 00042	0 16	0 06	0 00004
	<=3000	6918201	0 12	0 001226	0 00032	0 00141	0 37	0 15	0 00010
	>3000	185628296	1 71	0 018219	0 00289	0 01469	5 40	2 12	0 00137
	Total	194718163	1 88	0 020682	0 00364	0 01652	5 94	2 33	0 00151
Eastern Europe to Other	Ground	1678357	0 04	0 000670	0 00013	0 00048	0 14	0 05	0 00004
	<=3000	5346538	0 09	0 000543	0 00009	0 00143	0 30	0 12	0 00008
	>3000	238469415	1 92	0 009786	0 00113	0 02173	6 05	2 37	0 00153
	Total	245494310	2 06	0 010999	0 00134	0 02364	6 49	2 54	0 00164
Middle East to Middle East	Ground	2875884	0 09	0 001181	0 00026	0 00154	0 30	0 12	0 00008
	<=3000	9997934	0 22	0 001048	0 00020	0 00447	0 68	0 27	0 00017
	>3000	211823383	2 15	0 010717	0 00176	0 03206	6 78	2 66	0 00172
	Total	224697200	2 46	0 012945	0 00222	0 03808	7 76	3 04	0 00197
Middle East to Other	Ground	1306551	0 06	0 000813	0 00022	0 00107	0 19	0 07	0 00005
	<=3000	4522521	0 15	0 000676	0 00014	0 00338	0 49	0 19	0 00012
	>3000	383148043	4 86	0 016688	0 00231	0 06854	15 34	6 02	0 00389
	Total	388977115	5 08	0 018177	0 00267	0 07298	16 02	6 28	0 00406
North America and Caribbean to North America and Caribbean	Ground	127824922	2 10	0 035167	0 00582	0 01883	6 63	2 60	0 00168
	<=3000	302880697	3 72	0 017006	0 00284	0 05628	11 75	4 61	0 00298
	>3000	7927196149	57 11	0 157525	0 02217	0 71097	180 17	70 64	0 04568
	Total	8357901768	62 93	0 209698	0 03083	0 78608	198 54	77 84	0 05034
North America and Caribbean to Other	Ground	3309134	0 16	0 002023	0 00043	0 00264	0 49	0 19	0 00013
	<=3000	8744377	0 38	0 001226	0 00021	0 00881	1 19	0 47	0 00030
	>3000	1384542553	22 12	0 052276	0 00688	0 32905	69 78	27 36	0 01769
	Total	1396575664	22 65	0 055525	0 00753	0 34050	71 47	28 02	0 01812
South America to South America	Ground	6479637	0 14	0 001758	0 00023	0 00143	0 45	0 18	0 00011
	<=3000	22095560	0 29	0 001210	0 00016	0 00414	0 92	0 36	0 00023
	>3000	418968661	2 84	0 007876	0 00118	0 03229	8 94	3 51	0 00227
	Total	447543858	3 27	0 010844	0 00158	0 03787	10 31	4 04	0 00261
South America to Other	Ground	720532	0 03	0 000343	0 00009	0 00045	0 09	0 04	0 00002
	<=3000	2207158	0 07	0 000258	0 00006	0 00138	0 21	0 08	0 00005
	>3000	249606160	3 21	0 006324	0 00089	0 04473	10 13	3 97	0 00257
	Total	252533850	3 31	0 006925	0 00105	0 04656	10 43	4 09	0 00264
Western Europe and North Atlantic to Western Europe and North Atlantic	Ground	35536239	0 62	0 008448	0 00081	0 00592	1 94	0 76	0 00049
	<=3000	99001175	1 31	0 005030	0 00044	0 01922	4 12	1 62	0 00105
	>3000	1865526615	11 70	0 036219	0 00310	0 14207	36 92	14 48	0 00936
	Total	2000064029	13 62	0 049697	0 00435	0 16721	42 99	16 85	0 01090
Western Europe and North Atlantic to Other	Ground	4725818	0 19	0 002429	0 00048	0 00317	0 60	0 23	0 00015
	<=3000	13536790	0 46	0 001566	0 00025	0 01051	1 44	0 57	0 00037
	>3000	1490109091	21 96	0 054150	0 00697	0 32791	69 28	27 16	0 01757
	Total	1508371699	22 60	0 058145	0 00770	0 34159	71 32	27 96	0 01808

		Distance (nm)	Fuel Burn (Tg)	CO (Tg)	HC (Tg)	NOx (Tg)	CO2 (Tg)	H2O (Tg)	SOx (Tg)
Global Total	Ground	211742417	4 25	0 062978	0 01041	0 04821	13 42	5 26	0 00340
	<=3000	555959205	8 65	0 036048	0 00555	0 14911	27 30	10 70	0 00692

Figure 10. Example Regional Inventory

Country	Domestic/International	Mode	Distance (nm)	Fuel Burn (Kg)	CO (kg)	HC (kg)	NOx (kg)	CO2 (kg)	H2O (kg)	SOx (kg)
Australia	Domestic	Ground	2733794	49770128	594416	88902	667887	157024753	61505648	39816
Australia	Domestic	<=3000	8066560	102664631	381485	52661	1835714	323906912	126996149	82132
Australia	Domestic	>3000	202580225	1347062473	3122390	557511	18570644	4249982103	1666316279	1077650
Australia	Domestic	Total	213380580	1499497232	4098290	699074	21074245	4730913769	1854878077	1199598
Australia	International	Ground	319288	16427193	161674	28795	319873	51827794	20320438	13142
Australia	International	<=3000	985802	41952744	114974	17781	1016521	132360909	51855545	33562
Australia	International	>3000	125812540	2226335256	2720249	629976	35247016	7024087734	2753976712	1781068
Australia	International	Total	127117631	2284715194	2996897	676553	36583410	7208276437	2826192695	1827772
France	Domestic	Ground	3528887	50857338	817228	81042	441684	160454902	62910527	40686
France	Domestic	<=3000	8279867	99552459	383214	43739	1538971	314088008	123146392	79642
France	Domestic	>3000	124392267	692885843	1993035	259255	9155582	2186054834	857099787	554309
France	Domestic	Total	136201021	843295640	3193478	384035	11136237	2660597744	1043156706	674637
France	International	Ground	3463455	70684247	1070019	139538	872011	223008801	87435414	56547
France	International	<=3000	8028482	148856968	535043	64076	2853607	469643735	184156070	119086
France	International	>3000	355977803	4207764606	10213297	1288014	61134681	13275497332	5205004818	3366212
France	International	Total	367469740	4427305822	11818359	1491627	64860300	13968149868	5476577302	3541845
Germany	Domestic	Ground	2097676	42044221	426486	43321	489812	132649517	52003701	33635
Germany	Domestic	<=3000	5948782	81781667	246555	26397	1310960	258021161	101163923	65425
Germany	Domestic	>3000	69019106	445165571	1063838	124682	6058169	1404497378	550609812	356132
Germany	Domestic	Total	77065564	568991460	1736879	194401	7858941	1795168055	703842436	455193
Germany	International	Ground	3846090	90718018	1161270	163061	1154540	286215347	112218188	72574
Germany	International	<=3000	11219052	207949668	736377	83665	3733606	656081204	257233740	166360
Germany	International	>3000	491128247	5368608918	11518754	1353844	74063286	16937961137	6640969232	4294887
Germany	International	Total	506193389	5667276605	13416402	1600570	78951432	17880257689	7010421160	4533821
Ireland	Domestic	Ground	85730	1184880	17188	2083	13334	3738297	1465697	948
Ireland	Domestic	<=3000	234052	2485637	10793	1008	41438	7842184	3074733	1989
Ireland	Domestic	>3000	1455503	8859572	28226	2233	149515	27951948	10959290	7088
Ireland	Domestic	Total	1775285	12530089	56206	5324	204287	39532430	15499720	10024
Ireland	International	Ground	644392	14004343	171281	27965	166927	44183703	17322373	11203
Ireland	International	<=3000	1928377	32072275	109004	16138	521879	101188028	39675404	25658
Ireland	International	>3000	46828821	400407441	1024894	151787	5203158	1263285477	495304005	320326
Ireland	International	Total	49401590	446484060	1305180	195889	5891963	1408657208	552307782	357187
Mexico	Domestic	Ground	2316895	44405849	514365	97739	358351	140100455	54930036	35525
Mexico	Domestic	<=3000	7815480	99782687	340339	53236	1309612	314814378	123431184	79826
Mexico	Domestic	>3000	155627977	1032995415	3053806	413842	10651963	3259100534	1277815328	826396
Mexico	Domestic	Total	165760352	1177183951	3908510	564816	12319926	3714015367	1456176548	941747
Mexico	International	Ground	972674	23786810	330543	69768	241530	75047387	29424285	19029
Mexico	International	<=3000	2999260	54537069	238026	49368	809600	172064453	67462354	43630
Mexico	International	>3000	150138716	1346246866	4112982	540458	16246236	4247408863	1665307373	1076997
Mexico	International	Total	154110650	1424570746	4681551	659594	17297366	4494520703	1762194012	1139657
Peru	Domestic	Ground	249634	6434198	54595	13608	56431	20299894	7959103	5147
Peru	Domestic	<=3000	945982	14418948	44183	10284	176149	45491782	17836239	11535
Peru	Domestic	>3000	13360665	102419559	233844	56686	1047360	323133708	126692994	81936
Peru	Domestic	Total	14556282	123272705	332622	80577	1279940	388925384	152488336	98618
Peru	International	Ground	93241	3415702	38991	9834	44568	10776539	4225223	2733
Peru	International	<=3000	312945	7248775	29359	7079	132066	22869886	8966735	5799
Peru	International	>3000	27000020	267099604	816632	108933	3343677	842699252	330402211	213680
Peru	International	Total	27406205	277764082	884982	125846	3520311	876345678	343594169	222211
Sweden	Domestic	Ground	989929	13138183	171825	20285	127975	41450966	16251932	10511
Sweden	Domestic	<=3000	2829388	32689060	105944	9550	470290	103133986	40436368	26151
Sweden	Domestic	>3000	36017712	203318991	539375	58103	2637637	641471416	251505591	162655
Sweden	Domestic	Total	39837029	249146234	817144	87938	3235902	786056367	308193899	199317
Sweden	International	Ground	715501	14040880	174749	18719	142704	44298975	17368568	11233
Sweden	International	<=3000	2131886	32392499	108960	9691	481227	102198335	40069521	25914
Sweden	International	>3000	61827521	456825733	1212368	108097	5554762	1441285187	565093432	365461
Sweden	International	Total	64674908	503259112	1496077	136507	6178693	1587782497	622531521	402607
United Kingdom	Domestic	Ground	2797535	41423069	606463	61714	367779	130689784	51240037	33138
United Kingdom	Domestic	<=3000	7533961	79387431	330461	34825	1203189	250467346	98202253	63510
United Kingdom	Domestic	>3000	89131097	487805381	1298822	157361	6785275	1539025976	603415256	390244
United Kingdom	Domestic	Total	99462592	608615881	2235745	253899	8356243	1920183106	752857545	486893
United Kingdom	International	Ground	4473238	116194877	1524573	205367	1525285	366594835	143733362	92956
United Kingdom	International	<=3000	12662775	254233542	928260	117551	5015287	802106827	314486392	203387
United Kingdom	International	>3000	612018863	7958253389	24965599	3206134	121938214	25108289441	9844355442	6366603
United Kingdom	International	Total	629154876	8328681808	27418432	3529052	128478786	26276991103	10302573396	6662945
United States of America	Domestic	Ground	106997085	1779265360	30418108	4852571	15824556	5613582210	2200951250	1423412
United States of America	Domestic	<=3000	243992718	3080864237	14142979	2300600	46680074	9720126667	3811025061	2464691
United States of America	Domestic	>3000	6386831517	47225010595	126416635	17522500	594435828	148994908426	58417333106	37780008
United States of America	Domestic	Total	6737821320	52085140191	170977721	24675671	656940459	164328617303	64429318416	41668112

Figure 11. Example Country Inventory for Selected Countries

Aircraft	Engine	Distance (nm)	Fuel Burn (kg)	CO2 (kg)	H2O (kg)	SOx (kg)	CO (kg)	HC (kg)	NOx (kg)
A332	CF6-80E1A3	15335851	238432279	752253842	294940730	190746	303289	69635	4850917
	CF6-80E1A4	6884316	97033038	306139235	120029868	77626	65966	7874	1511083
	PW4168A	35184618	486694736	1535521893	602041389	389356	730837	52589	9328606
	TRENT772B-60	24657181	349845979	1103764063	432759476	279877	235554	5918	5078192
A333	CF6-80E1A2	5260471	62395668	196858334	77183442	49917	91332	5770	862084
	PW4168	5687363	79120307	249624570	97871820	63296	171660	222	1149907
	PW4168A	14337554	174391217	550204288	215721935	139513	318916	26344	3081316
	TRENT768-60	40326	524304	1654180	648564	419	492	14	7436
	TRENT772-60	418461	5474321	17271481	6771735	4379	5416	155	79679
	TRENT772B-60	10167356	125114440	394736058	154766562	100092	117412	2655	1675562
A340	CFM56-5C2	59945188	931763182	2939712839	1152591056	745411	1585759	42329	15663537
	CFM56-5C3\F	190758738	2971831710	9376129044	3676155825	2377465	5203313	135561	50539602
A342	CFM56-5C2	4626317	72099751	227474713	89187392	57680	120450	2931	1209068
	CFM56-5C3\F	600272	9209885	29057188	11392628	7368	15639	318	153691
A343	CFM56-5C2	21518487	336265117	1060916444	415959950	269012	571538	15110	5670546
	CFM56-5C3	3383894	52652649	166119106	65131326	42122	92907	2346	895940
	CFM56-5C3\F	30827043	481257576	1518367654	595315622	385006	843272	20812	8200294
	CFM56-5C4	85537295	1332941022	4205428925	1648848045	1066353	2378914	58418	22955430
B733	CFM56-3B1	565844610	3914746188	12351024224	4842541035	3131797	14029821	690353	45235918
	CFM56-3B2	332197764	2290931770	7227889735	2833882600	1832745	8297045	358119	26652371
	CFM56-3C1	167264603	1037313983	3272725615	1283157397	829851	3322274	111403	11311081
B734	CFM56-3B1	61578768	432016330	1363011523	534404201	345613	1347599	68156	5103241
	CFM56-3B2	63302435	473403106	1493586800	585599642	378722	1662350	75553	5810925
	CFM56-3C1	199335815	1452329725	4582100282	1796531870	1161864	4793697	187863	17950250
B735	CFM56-3B1	151039192	1069434220	3374064963	1322890130	855547	4331509	230760	12388755
	CFM56-3B2	30812935	217310472	685614538	268813053	173848	843364	40631	2509486
	CFM56-3C1	119912842	839003656	2647056534	1037847522	671203	2977654	124577	9659452
B736	CFM56-7B20	24407101	171606063	541417130	212276700	137285	477762	47389	2241396
B741	JT9D-3A	439508936	10529683965	33221152908	13025219064	8423747	11556979	5943936	175957826
	JT9D-7	42427685	1008158411	3180739788	1247091955	806527	921442	538022	16800423
	JT9D-7A	48599062	1181255022	3726859594	1461212462	945004	1713437	794400	20189907
	JT9D-7AH	10270711	246447120	777540663	304855087	197158	352059	167313	4169819
	JT9D-7F	46418867	1108810867	3498298285	1371599042	887049	2973409	273757	20204839
	JT9D-7R4G2	44575204	1084681548	3422170283	1341751074	867745	1393615	282673	18321050
	RB211-524C2	109019	2615185	8250910	3234984	2092	32103	5279	40960
B742	CF6-50E2	19940654	528588148	1667695606	653863539	422871	638849	141031	8417669
	JT9D-7	7634	200100	631317	247524	160	115	79	3688
	JT9D-70A	502707	13798821	43535280	17069141	11039	13975	6426	205732
	JT9D-7A	3607575	97995133	309174646	121219980	78396	93405	52764	1859368
	JT9D-7AW	908376	25300244	79822271	31296402	20240	28468	14751	488208
	JT9D-7F	1865885	49453491	156025764	61173968	39563	111286	8727	989983
	JT9D-7FW	9918	279147	880708	345304	223	652	71	6030
	JT9D-7J	2323056	63032801	198868489	77971575	50426	164022	12599	1268713
	JT9D-7Q	17659048	473031078	1492413050	585139443	378425	404655	207179	6972205
	JT9D-7Q3	2197032	57707645	182067621	71384357	46166	44387	24586	845396
	JT9D-7R4G2	6446630	171855859	542205234	212585697	137485	218478	44597	3160126
	JT9D-7W	81180	2121948	6694746	2624850	1698	1537	881	30648
	RB211-524B2	26295	669731	2113001	828457	536	624	399	15640
	RB211-524C2	675894	18048697	56943640	22326238	14439	141790	23465	312998
	RB211-524D4	7536041	202743815	639656736	250794099	162195	238477	144690	4958894
B743	CF6-50E2	14484847	341093113	1076148771	421932181	272874	744814	90171	4902877
	CF6-80C2B1	8791222	215842361	680982650	266997001	172674	120499	21058	3017013
	JT9D-7R4G2	18278190	444360192	1401956406	549673557	355488	555028	114243	7385388
	RB211-524C2	10098796	240712607	759448276	297761495	192570	3241629	564085	3712442
	RB211-524D4	19165242	454758635	1434763492	562536431	363807	543261	344426	9838013
B744	CF6-80C2B1F	421027542	10147986178	32016896392	12553058903	8118389	7717286	1010028	137359794
	PW4056	342665948	8210931120	25905487684	10156921796	6568745	9907253	584788	119846909
	RB211-524G	18884869	446270658	1407983926	552036804	357017	430972	253208	8322185
	RB211-524G\H-T	3929738	95064803	299929452	117595161	76052	68327	1335	1191629
	RB211-524H	4542089	108908600	343606633	134719938	87127	121111	64151	2213176

Figure 12. Example Aircraft-Engine Inventory for Selected Equipment

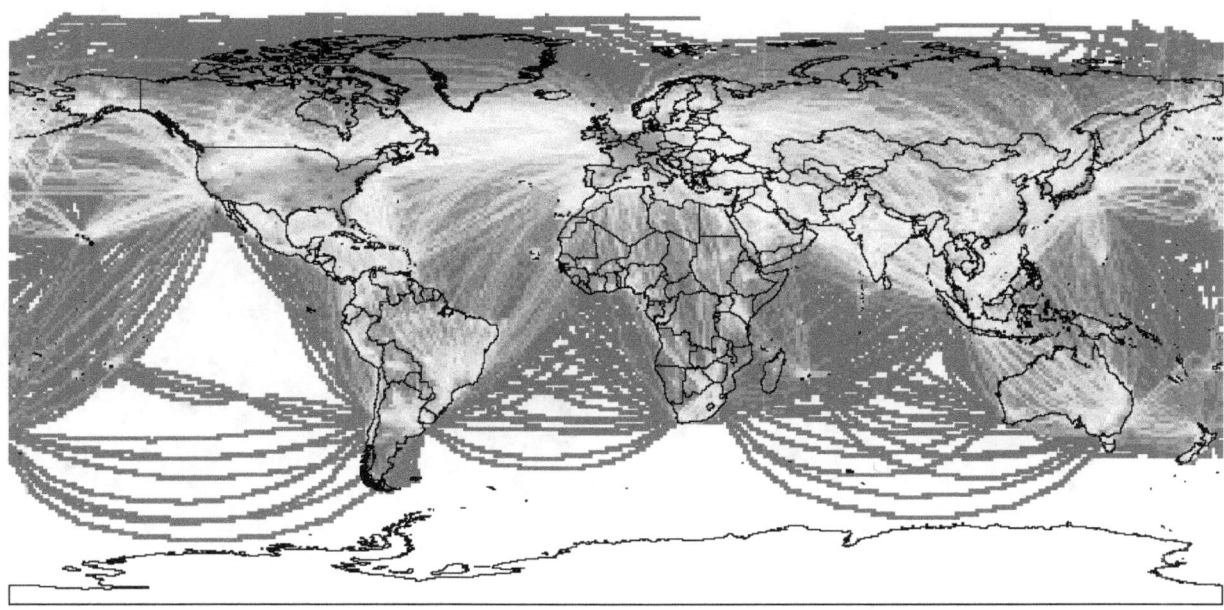

Figure 13. Example Fuel Burn Plot of 1° Latitude by 1° Longitude World Grids with all Altitudes aggregated

The country and regional inventories exemplified in figures 9 and 10 are based on airport locations and the origin airport. The definitions for domestic (e.g., region A to region A) and international (e.g., region A to any other region) flights are described in Appendix A. For illustration purposes, the Figure 13 shows locations of all worldwide airports color-coded to reflect the eight different regions.

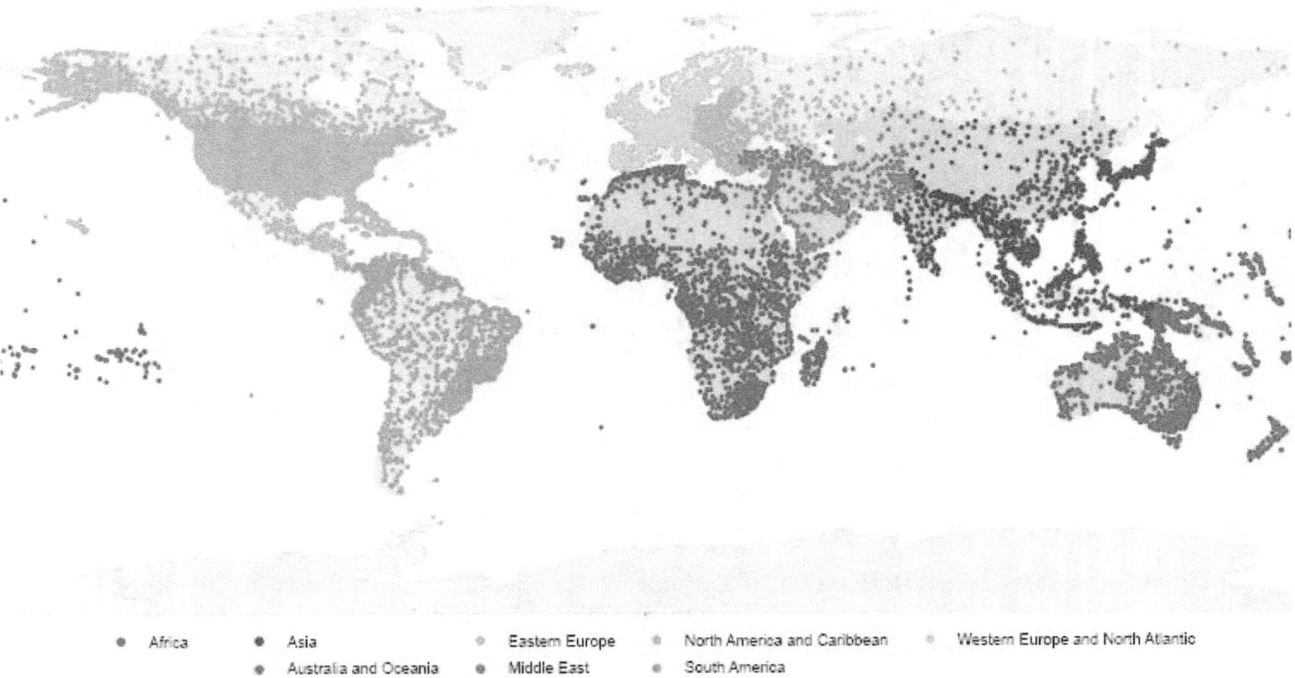

Figure 14. Worldwide Airport Locations Color-Coded by Region

All of the exemplified processed inventories show the usefulness of the raw inventories for conducting various scenario analyses within SAGE. The regional and country inventories could potentially be used to determine the relative effects of emissions stringency policies on various regions/countries. The aircraft-engine inventory could potentially be used to help target the worst polluters in an aircraft phase-out assessment. And the aggregated grid inventories could be provided to atmospheric dispersion modelers (possibly in finer geometric and time slices) to conduct fate studies of the various pollutants.

Because SAGE can and has been used to generate inventories for various historical years (e.g., 2000-2004), the results can be used to study trends in various metrics such as fuel per distance, emissions per operations, loadings by altitude, etc. These types of studies would help to determine the effectiveness of national policies for certain countries and regions. Combining these querying capabilities with the ability to dynamically model aircraft performance, trajectories, capacity/delay, forecasting parameters, etc. provides the user with a powerful set of modeling options.

Since the modeling capabilities within SAGE would be meaningless if the results were untrustworthy, FAA has invested significant resources to conduct validation assessments of the model. On a system level, the model has shown to produce good overall agreement (less than 3% difference) with measured fuel burn data from several major airlines. Further details of the validation assessments can be found in FAA[d] 2005.

6 CONCLUSIONS

SAGE was developed by FAA in large part because there was no up-to-date non-proprietary model that could be used to estimate aircraft emissions on a global level. As such, FAA developed SAGE, now at Version 1.5, from the best publicly available data and methods in order to provide the international aviation community with a high-fidelity tool that can be used to analyze various policy, technology, and operational scenarios. So far, SAGE has been used to develop inventories for 2000-2004. The current commitment from FAA is to continue development and validation of SAGE to produce inventories of fuel burn and emissions on a yearly basis.

The primary elements within SAGE include aircraft movements data, aircraft performance (including fuel flow), emissions, capacity and delay, and forecasting. All of these components reside within a dynamic modeling environment where the user has access to all of the input parameters that can be queried and modified to evaluate various what-if scenarios

The movements database within SAGE was developed by combining ETMS radar data with OAG simulated data. This difficult combination represents a concerted effort to provide the most accurate coverage of global aircraft movements. The database includes all flights in a given year such that no approximations are needed to scale a smaller representative set of flights. Aircraft performance is modeled through Eurocontrol's BADA data and methodology. The BADA database represents one of the most comprehensive set of publicly available aircraft performance data that can be used to model all modes of flight. With the continuing support from Eurocontrol and growing list of users, BADA is poised to become the standard in publicly available aircraft performance data and methodology. Emissions modeling is conducted largely through the use of BFFM2 which has been internationally recognized as an acceptable methodology by ICAO CAEP. Partly due to these types of recognitions and the increased use by the international community in recent years, BFFM2 is the de facto standard in modeling aircraft emissions for CO, HC, and NOx. And Boeing's continued support of BFFM2 will help to ensure that the method is continuously improved and stays relevant in the future. Capacity and delay modeling is conducted through the use of WWLMINET, a sub-model within SAGE. WWLMINET and its predecessor, LMINET, are internationally recognized models that allow for efficient determination of terminal area delays and provide various airport capacity-related scenarios to be modeled including those related to future years. Forecasting is conducted through the use of internationally recognized standard datasets including FAA's TAF and ICAO CAEP FESG's forecasts. Although forecasting is by nature a dubious science, these datasets represent inputs from various international experts, and therefore, provide credibility to the modeled results.

In addition to these modeling capabilities, the raw output inventories provide a high-resolution coverage of flight modeling parameters and fuel burn/emissions results that can be used for various analyses including but not limited to studies concerning fuel efficiency, emissions stringency, aircraft phase-outs, and atmospheric dispersion modeling.

All of the capabilities and the concerted effort to implement the best publicly available data and methods within SAGE are based on FAA's goal of providing the international aviation community with a standard and open model that can provide high-fidelity results in helping to answer various policy, technology, and operations questions. Again, FAA is committed to supporting the continued development and validation of SAGE in producing yearly inventories of fuel burn and emissions. Therefore, the data and methods within SAGE will stay relevant year after year.

REFERENCES

BACK Aviation Solutions (BACK). "Aviation Link: FLEET. User's Guide." Edition 16. June 2005.

Baughcum[a], S.L., T.G. Tritz, S.C. Henderson, and D.C. Pickett. "Scheduled Civil Aircraft Emission Inventories for 1992: Database Development and Analysis." NASA CR 4700. April 1996.

Baughcum[b], S. L., S. C. Henderson, and T. G. Tritz. "Scheduled Civil Aircraft Emission Inventories for 1976 and 1984: Database Development and Analysis." NASA CR-4722. 1996.

Baughcum[c], S. L., S. C. Henderson, and D. J. Sutkus. "Scheduled Civil Aircraft Emission Inventories Projected for 2015: Database Development and Analysis." NASA CR-1998-207638. 1998.

Baughcum[d], S. L., and S. C. Henderson. "Aircraft Emission Scenarios Projected in Year 2015 for the NASA Technology Concept Aircraft (TCA) High Speed Civil Transport." NASA/CR-1998-207635. 1998.

Bureau of Transportation Statistics (BTS[a]). "Airline On-Time Performance Data." http://www.transtats.bts.gov/Tables.asp?DB_ID=120&DB_Name=Airline%20On-Time%20Performance%20Data&DB_Short_Name=On-Time. 2005.

Bureau of Transportation Statistics (BTS[b]). "Aviation Support Tables. Master Coordinate." http://www.transtats.bts.gov/Tables.asp?DB_ID=595&DB_Name=Aviation%20Support%20Tables&DB_Short_Name=Aviation%20Support%20Tables. 2005.

Daggett, D. L., D. J. Sutkus Jr., D. P. DuBois, and S. L. Baughcum. "An Evaluation of Aircraft Emission inventory Methodology by Comparisons with Reported Airline Data." 1999-CR-209480. 1999.

Environmental Systems Research Institute (ESRI). "ArcGIS – The comprehensive Geographic Information System." http://www.esri.com/software/arcgis/. 2005.

Eurocontrol Experimental Center (EEC). "User Manaul for the Base of Aircraft Data (BADA), Revision 3.6." EEC Note No. 10/04. Project ACE-C-E2. September 2004.

Eurocontrol. "European Database of Major Airports in the ECAC States." 1998.

Federal Aviation Administration (FAA). "Emissions and Dispersion Modeling System (EDMS) Reference Manual." CSSI, Inc. Washington, DC. May 2001.

Federal Aviation Administration (FAA). "Integrated Noise Model (INM) Version 6.0 Technical Manual." US Departement of Transporation/Volpe National Transportation Systems Center. ATAC Corporation. OMB No. 0704-0188. FAA-AEE-02-01. January 2002.

Federal Aviation Administration (FAA). "Airport Capacity Benchmark Report 2001." http://www.mitrecaasd.org/library/general/airport_capacity_benchmarks.pdf. May 28, 2004.

Federal Aviation Administration (FAA[a]). "Official Airline Guide." FAA Office of Aviation Policy and Plans (APO). http://apo.faa.gov/apo_home.asp. 2005.

Federal Aviation Administration (FAA[b]). "Terminal Area Forecast." Federal Aviation Administration, Office of Aviation Policy and Plans (APO). http://www.apo.data.faa.gov/faatafall.HTM. 2005.

Federal Aviation Administration (FAA[c]). "System for assessing Aviation's Global Emissions (SAGE), Version 1.5, Global Aviation Emissions Inventories for 2000 through 2004." FAA, Office of Environment and Energy. FAA-AEE-2005-02. September 2005.

Federal Aviation Administration (FAA[d]). "System for assessing Aviation's Global Emissions (SAGE), Version 1.5, Validation Assessment, Model Assumptions, and Uncertainties." FAA, Office of Environment and Energy. FAA-AEE-2005-03. September 2005.

Flathers, G. W., II. A Comparison of FAA Integrated Noise Flight Profiles with Profiles Observed at Seattle-Tacoma Airport. FAA-EE-82-10, December 1982.

Forecasting and Economics Sub Group (FESG). Steering Group Meeting Report of the FESG/CAEP6 Traffic and Fleet Forecast. FESG CAEP-SG20031-IP/8. October 1003.

Gardner, R. "Global Aircraft Emissions Inventories for 1991/92 and 2015, Report by the ECAC/ANCAT and ED Working Group." Editor: R. M. Gardner. EUR18179. 1998.

Grandi, Fabio. "The MAGENTA Modeling System – Software and Data Structures – Data Management Software Utilities." Wyle Laboratories. Washington, DC. 2005.

Hadaller, O. J., and A. M. Momenthy. "The Characteristics of Future Fuels." Boeing publication D6-54940. 1989.

Hadaller, O. J., and A. M. Momenthy. "Characteristics of Future Aviation Fuels." Chapter 10 in Transportation and Global Climate Change, (D. L. Greene and D. J. Santini, Eds.). American Council for an Energy-Efficient Economy. Washington, DC. 1993.

International Civil Aviation Organization (ICAO). "ICAO Engine Exhaust Emissions Databank, First Edition." Doc 9646-AN/943. 1995.

International Civil Aviation Organization (ICAO[a]). "ICAO Engine Exhaust Emissions Databank, Issue 13. http://www.caa.co.uk/default.aspx?categoryid=702&pagetype=90. October 2004.

International Civil Avitaion Organization (ICAO[b]). "Guidance on the use of LTO Emissions Certification Data for the Assessment of Operational Impacts." ICAO Committee on Aviation Environmetnal Protection (CAEP) Working Group 3 (WG3), Alternative Emissions Task Group (AEMTG). CAEP/6-IP/5. Montreal, Canada. February 2004.

International Civil Aviation Organization (ICAO). "Issues with Implementing Boeing Fuel Flow Method 2 (BFFM2) for Computing Emissions using the ICAO Databank." ICAO Committee on Aviation Environmetnal Protection (CAEP) Working Group 3 (WG3), Alternative Emissions Task Group (AEMTG). IP-4/3. Stockholm, Sweden. June 20-25, 2005.

Intergovernmental Panel on Climate Change (IPCC). "Aviation and the Global Atmosphere." A Special Report of IPCC Working Groups I and II. Edited by J.E. Penner, D.H. Lister, D.J. Griggs, D.J. Dokken, and M. McFarland. Cambridge University Press. 1999.

Jane's Information Group, Inc. (Jane's). "Jane's All the World's Aircraft 2004-2005." Editor in Chief: Paul Jackson. Deputy Editor: Kenneth Munson. Alexandria, Virginia 22314. 2005.

JP Fleets. "JP Airline Fleets International, Aviation Database CD-Rom." BUCHair UK Ltd. December 2004.

Klima, Kelly. "Assessment of a Global Contrail Modeling Method and Operational Strategies for Contrail Mitigation." Master's Thesis in the Department of Aeronautics and Astronautics, Massachusetts Institute of Technology (MIT). May 2005.

Lee, Joosung. "Modeling Aviation's Global Emissions, Uncertainty Analysis, and Applications to Policy." Ph.D Thesis in the Department of Aeronautics and Astronautics, Massachusetts Institute of Technology (MIT). February 2005.

Long, Dou, Earl Wingrove, David Lee, Joana Gribko, Robert Hemm, and Peter Kostiuk. "A Method for Evaluating Air Carrier Operational Strategies and Forecasting Air Traffic with Flight Delay." National Aeronautics and Space Administration (NASA). NASA Contract No. NAS2-14361. NS902S1. Logistics Management Institute (LMI). McLean, Virginia 22102-7805. October 1999.

Lundgren, Johan. www.airliners.net. Luleå University of Technology. Sponsored by AirNav Systems LLC. 1995-2005.

Michot, Sophie, Ted Elliff, Gregg G. Fleming, Brian Kim, Curtis A. Holsclaw, Maryalice Locke, and Angel Morales. "Flight Movement Inventory: SAGE-AERO2K." Air Traffic Control Quarterly. Vol 12(2) 125-145. 2004.

National Aeuronautics and Space Administration (NASA). "Modeling Air Traffic Management Technologies with a Queuring Network Model of the National Airspace System." NASA Contractor Report 208988. Dou Long, David A. Lee, Jesse P. Johnson, Eric M. Gaier, and Peter F. Kostiuk. 1998.

QinetiQ (2004). "ICAO Engine Exhaust Emissions Databank." Hosted by QinetiQ on the internet at: http://www.qinetiq.com/aviation_emissions_databank. February 2004.

Schmitt, A., B. Brunner. "Emissions from Aviation and their Development over Time. In Final Report on the BMBF Verbundprogramm, Schadstoff in der Luftfahrt. DLR-Mitteilung 97-04, Deutches Centrum Fuer Luft- and Raumfahrt. 1997.

Society of Automotive Engineers (SAE). Society of automotive Engineers Aerospace Information Report 1845. Warrandale, PA. 1986.

Sutkus Jr., Donald J., Steven L. Baughcum, and Douglas P. DuBois. "Scheduled Civil Aircraft Emission Inventories for 1999: Database Development and Analysis." National Aeronatics and Space Administration (NASA) Glenn Research Center, Contract NAS1-20341. NASA/CR-2001-211216. October 2001.

United States Department of Transportation (USDOT). "FAA Aerospace Forecasts—Fiscal Years 1999–2010." Report No. FAA APO-99-1. Federal Aviation Administration, Office of Aviation Policy and Plans, Statistics and Forecast Branch. Washington, D.C. March 1999.

United States Geological Survey (USGS). "GTOPO30 Global Digital Elevation Model." http://edcdaac.usgs.gov/gtopo30/gtopo30.asp. http://www.usgs.gov. 2005.

Volpe National Transportation Systems Center/US DOT. "Enhanced Traffic Management System (ETMS), Functional Description, Version 7.6." Report Number VNTSC-DTS56-TMS-002. August 2003.

APPENDIX A: Country-Region Mapping Tables

The following list shows all of the countries and corresponding regions implemented within SAGE. Flight results (fuel burn and emissions) are attributed to these countries and regions based on location of airports. A domestic flight for a country or region is defined as a flight that departs and arrives within the same country or region. In contrast, an international flight is defined as a flight that departs from one country or region and arrives in another country or region.

Africa		
Algeria	Ghana	Seychelles
Angola	Guinea	Sierra Leone
Benin	Guinea-Bissau	Somalia
Botswana	Kenya	South Africa
Burkina Faso	Lesotho	Sudan
Burundi	Liberia	Swaziland
Cameroon	Libya	Tanzania
Cape Verde	Madagascar	Togo
Central African Republic	Malawi	Tunisia
Chad	Mali	Uganda
Comoros	Mauritania	Zambia
Congo, Republic of	Mauritius	Zimbabwe
Congo, Democratic Republic of	Morocco	*Territories*:
Cote d'Ivoire (Ivory Coast)	Mozambique	France: Department of Réunion (island)
Djibouti	Namibia	UK: British Indian Ocean Territory, St. Helena (including Ascension)
Equatorial Guinea	Niger	
Eritrea	Nigeria	
Ethiopia	Rwanda	
Gabon	Sao Tome and Principe	
Gambia	Senegal	

Eastern Europe		
Albania	Estonia	Poland
Belarus	Greece	Romania
Bosnia and Herzegovina	Hungary	Russia
Bulgaria	Latvia	Serbia and Montenegro
Croatia	Lithuania	Slovakia (Slovak Republic)
Cyprus	Macedonia	Slovenia
Czech Republic	Moldova	Ukraine

Western Europe & North Atlantic		
Andorra	Iceland	Norway
Austria	Ireland	Portugal
Belgium	Italy	San Marino
Denmark	Liechtenstein	Spain
Finland	Luxembourg	Sweden

France Germany Greenland (Kalaallit Nunaat)	Malta Monaco Netherlands	Switzerland United Kingdom

South America		
Argentina Bolivia Brazil Chile Colombia Ecuador	Guyana Panama Paraguay Peru Suriname Uruguay	Venezuela <u>& Other Territories</u>: Chile: Easter Island France: French Guiana UK: Falkland Islands

North America & Caribbean		
Antigua and Barbuda The Bahamas Barbados Belize Canada Costa Rica Cuba Dominica Dominican Republic El Salvador Grenada Guatemala	Haiti Honduras Jamaica Mexico Nicaragua Saint Kitts and Nevis Saint Lucia Saint Vincent and the Grenadines Trinidad and Tobago United States of America	<u>& Other Territories</u>: France: French Antilles (Martinique, Guadeloupe, St. Martin, St. Barthelemy) Netherlands: Aruba, Netherlands Antilles (Curaçao, Bonaire, St. Maarten, Saba, St. Eustatius) UK: Anguilla, Bermuda, British Virgin Islands, Cayman Islands, Montserrat, Turks and Caicos Islands US: Puerto Rico*, Virgin Islands* *Confirmed with EPA that these are *not* included in the annual US GHG Inventory total

Middle East		
Afghanistan Armenia Azerbaijan Bahrain Egypt Georgia Iran Iraq	Israel Jordan Kazakstan Kuwait Kyrgyzstan Lebanon Oman Pakistan Qatar	Saudi Arabia Syria Tajikistan Turkey Turkmenistan United Arab Emirates Uzbekistan Yemen

Asia		
Bangladesh	India	Mongolia
Bhutan	Indonesia	Nepal
Brunei	Japan	Philippines
Burma (Myanmar)	Korea, North	Singapore
Cambodia	Korea, South	Sri Lanka
China	Laos	Taiwan
East Timor	Malaysia	Thailand
Hong Kong	Maldives	Vietnam

Australia and Oceania		
Australia Fiji Kiribati Marshall Islands Micronesia Nauru New Zealand	Palau Papua New Guinea Samoa Solomon Islands Tonga Tuvalu Vanuatu	<u>& Other Territories:</u> France: French Polynesia (Marquesas, de la Société, etc.), New Caledonia, Wallis and Futuna Islands New Zealand: Niue United Kingdom: Pitcairn Island United States: American Samoa, Guam, Johnston Island (Atoll), Kingman Reef, Midway, Northern Mariana Islands, Palmyra, Wake Island

APPENDIX B: Analysis of Duplicate Flights in OAG

In an effort to quantify the effects of duplicate flights in OAG especially due to code-shares, the equipment, departure/arrival airports and dates/times were used to identify duplicates. The results of analyzing five years worth of data are shown below:

Year	Total Flights	Total Distance (nm)	Duplicate Flights	Duplicate Distance (nm)	% of Total Flights due to Duplicates	% of Total Distance due to Duplicates
2000	27,836,274	1,395,194,402	145,697	4,355,708	**0.52**	**0.31**
2001	27,512,578	1,363,983,652	83,881	2,194,130	**0.30**	**0.16**
2002	26,221,732	1,505,538,752	100,514	4,427,415	**0.38**	**0.29**
2003	26,009,265	1,666,039,089	86,001	3,075,349	**0.33**	**0.18**
2004	27,158,333	2,134,979,841	59,437	2,841,651	**0.22**	**0.13**
Aggregate	134,738,182	8,065,735,736	475,530	16,894,253	**0.35**	**0.21**

The distances are based on rough estimates rather than on actual or simulated trajectories. This was deemed acceptable for this analysis since the computed % values are relative. These results indicate that duplicate flights are essentially neglibible.

APPENDIX C: Helicopters Currently Removed from the SAGE Movements Database

The following equipment types have been identified as helicopters in ETMS and OAG for years 2000-2004, and are currently not modeled:

- A109
- AH1
- AS32
- AS35
- AS50
- AS65
- ASTR
- CH46
- CH53
- H500
- R22
- R44
- B109
- B206
- B212
- B222
- B412
- S61
- S65C
- S76
- UH1

This is a dynamic list that will continue to grow as more historical movements data are processed each year and additional helicopter flights are identified.

APPENDIX D: Aircraft Code Mappings used for Aircraft Performance Modeling

The following table presents mappings of ETMS and OAG aircraft codes to BADA and INM aircraft codes. These mappings are necessary in order to model aircraft performance within SAGE, and the mappings range from exact matches to approximate substitutions. The default engines are generally based on popularity within the BACK fleet database.

ETMS or OAG Aircraft Code	BADA Aircraft Code	INM Aircraft Code	BACK Default Engine
300	A306	A30062	PW4158
306	A306	A30062	PW4158
310	A310	A310	CF6-80C2A2
318	A319	A319	CFM56-5B6\P
319	A319	A319	CFM56-5A5
320	A320	A320	CFM56-5A1
330	A333	A330	TRENT772B-60
340	A343	A340	CFM56-5C4
342	A343	A340	CFM56-5C2
343	A343	A340	CFM56-5C4
703	B703	707320	JT3D-1
707	B703	707320	JT3D-1
712	B712	717200	BR715A1-30
717	B712	717200	BR715A1-30
721	B722	727100	JT8D-7B
722	B722	727100	JT8D-15
723Q	B722	727100	JT8D-15
727	B722	727100	JT8D-15
727Q	B722	727100	JT8D-15
732	B732	737D17	JT8D-15
733	B733	7373B2	CFM56-3B1
734	B734	737400	CFM56-3C1
735	B735	737500	CFM56-3C1
736	B736	737500	CFM56-7B20
737	B737	737700	CFM56-7B24
738	B738	737700	CFM56-7B26
741	B742	74710Q	JT9D-7A
742	B742	747200	CF6-50E2
743	B743	74720A	JT9D-7R4G2
744	B744	747400	CF6-80C2B1F
747	B744	747400	CF6-80C2B1F
752	B752	757RR	RB211-535E4
753	B753	757RR	RB211-535E4-C
757	B752	757RR	RB211-535E4
762	B762	767CF6	CF6-80A
763	B763	767CF6	PW4060
764	B763	767CF6	PW4060
765	B763	767CF6	PW4060
767	B763	767CF6	PW4060
772	B772	777200	GE90-90B
777	B772	777200	GE90-90B
A124	B742	74720B	D-18T
A125	B742	74720B	D-18T
A140	AT72	HS748A	TV-3-117VMA
A145	AT72	HS748A	TV-3-117VMA
A146	AT72	HS748A	TV-3-117VMA
A225	B744	747400	D-18T
A300	A306	A30062	PW4158
A301	A306	A30062	PW4158
A302	A306	A30062	PW4158
A303	A306	A30062	PW4158
A304	A306	A30062	PW4158
A305	A306	A30062	PW4158
A306	A306	A30062	CF6-80C2A5
A308	A306	A30062	PW4158
A309	A306	A30062	PW4158
A30A	A30B	A300	PW4158
A30B	A30B	A300	PW4158
A30G	A30B	A300	PW4158
A310	A310	A310	CF6-80C2A2
A311	A310	A310	CF6-80C2A2
A312	A310	A310	CF6-80C2A2
A313	A310	A310	CF6-80C2A2
A314	A310	A310	CF6-80C2A2
A315	A310	A310	CF6-80C2A2
A316	A319	A319	CFM56-5A5
A317	A319	A319	CFM56-5A5
A318	A319	A319	CFM56-5B8/P
A319	A319	A319	CFM56-5A5

A31O	A310	A310	CF6-80C2A2
A320	A320	A320	CFM56-5A1
A321	A321	A32123	CFM56-5B3\P
A322	A321	A32123	CFM56-5B3\P
A323	A321	A32123	CFM56-5B3\P
A324	A321	A32123	CFM56-5B3\P
A325	A321	A32123	CFM56-5B3\P
A326	A321	A32123	CFM56-5B3\P
A328	A321	A32123	CFM56-5B3\P
A329	A321	A32123	CFM56-5B3\P
A32O	A320	A320	CFM56-5A1
A330	A333	A330	TRENT772B-60
A331	A332	A33034	PW4168A
A332	A332	A33034	TRENT772B-60
A333	A333	A33034	PW4168
A334	A333	A33034	PW4168
A335	A333	A33034	PW4168
A336	A333	A33034	PW4168
A33A	A333	A33034	PW4168
A33O	A333	A330	TRENT772B-60
A340	A343	A340	CFM56-5C2
A341	A343	A340	CFM56-5C2
A342	A343	A340	CFM56-5C2
A343	A343	A340	CFM56-5C4
A344	A343	A340	CFM56-5C4
A345	A343	A340	TRENT553
A346	A343	A340	TRENT556-61
A348	A343	A340	TRENT556
A349	A343	A340	TRENT556
A34O	A343	A340	CFM56-5C4
A3O6	A306	A30062	PW4158
A3OB	A30B	A300	PW4158
A40	AT72	HS748A	PW127
A4F	B742	74720B	D-18T
A748	AT72	HS748A	DART7MK534-2
ABX	A30B	A30062	PW4158
ABY	A306	A30062	CF6-80C2A5
AC10	P28A	CNA172	CJ610-1
AC11	P28A	CNA172	CJ610-1
AC12	P28A	CNA172	CJ610-1
AC14	P28A	CNA172	CJ610-1
AC50	PA31	BEC58P	IO-540-K1B5
AC51	PA31	BEC58P	IO-540-K1B5
AC52	PA31	BEC58P	IO-540-K1B5
AC55	PA31	BEC58P	IO-540-K1B5
AC56	PA31	BEC58P	IO-540-K1B5
AC58	PA31	BEC58P	IO-540-K1B5
AC59	PA31	BEC58P	IO-540-K1B5
AC5O	PA31	BEC58P	IO-540-K1B5
AC90	PAY2	DHC6	PT6A-34
AC95	PAY2	DHC6	PT6A-34
AEST	PA31	BEC58P	IO-540-K1B5
AN06	AT72	HS748A	AI-24UT
AN12	C130	C130	AI-20M
AN22	C130	C130	NK-12MA
AN24	AT72	HS748A	AI-24UT
AN26	AT72	HS748A	AI-24UT
AN28	SH36	SD330	TWD-10B
AN30	AT72	HS748A	AI-24UT
AN32	AT72	HS748A	AI-20D
AN38	SH36	SD330	TWD-10B
AN6	AT72	HS748A	AI-24UT
AN72	DC94	DC9Q9	D-36
AN74	DC94	DC9Q9	D-36-2A
AN8	DH8C	DHC830	AI-20
ARJ	B462	BAE146	LF507-1H
ARJ1	B462	BAE146	LF507-1H
ARJ2	B462	BAE146	LF507-1H
ARVA	PAY2	DHC6	PT6A-36
ASTR	H25B	IA1125	TFE731-3-1G
AT42	AT43	DHC8	PW120
AT43	AT43	DHC8	PW120
AT44	AT43	DHC8	PW127E
AT45	AT45	DHC8	PW127E
AT72	AT72	HS748A	PW127
ATP	ATP	HS748A	PW126
B11	BA11	BAC111	SPEYMK506
B111	BA11	BAC111	SPEYMK506
B112	BA11	BAC111	SPEYMK506
B12	BA11	BAC111	SPEYMK506
B190	D228	1900D	PT6A-67D
B350	PAY2	DHC6	PT6A-60A
B46	B462	BAE146	ALF502R-5
B461	B462	BAE146	ALF502R-5
B462	B462	BAE146	ALF502R-5
B463	B462	BAE146	ALF502R-5
B464	B462	BAE146	ALF502R-5
B700	B703	707320	JT3D-1
B701	B703	707320	JT3D-1
B702	B703	707320	JT3D-1

B703	B703	707320	JT3D-1
B706	B703	707320	JT3D-1
B707	B703	707320	JT3D-1
B708	B703	707320	JT3D-1
B711	B712	717200	BR715A1-30
B712	B712	717200	BR715A1-30
B713	B712	717200	BR715A1-30
B714	B712	717200	BR715A1-30
B717	B712	717200	BR715A1-30
B719	B712	717200	BR715A1-30
B720	B703	707320	JT8D-7B
B721	B722	727100	JT8D-7B
B722	B722	727100	JT8D-15
B723	B722	727100	JT8D-15
B724	B722	727100	JT8D-15
B725	B722	727100	JT8D-15
B726	B722	727100	JT8D-15
B727	B722	727100	JT8D-15
B728	B722	727100	JT8D-15
B729	B722	727100	JT8D-15
B72A	B722	727100	JT8D-15
B72B	B722	727100	JT8D-15
B72C	B722	727100	JT8D-15
B72E	B722	717200	JT8D-15
B72F	B722	727100	JT8D-15
B72G	B722	727100	JT8D-15
B72I	B722	727100	JT8D-7B
B72K	B722	727100	JT8D-15
B72O	B722	727100	JT8D-7B
B72Q	B722	727100	JT8D-15
B72R	B722	727100	JT8D-15
B72S	B722	727100	JT8D-15
B72T	B722	727100	JT8D-15
B72W	B722	727100	JT8D-15
B72X	B722	727100	JT8D-15
B72Z	B722	727100	JT8D-15
B730	B732	7373B2	JT8D-7A
B731	B732	7373B2	JT8D-7A
B732	B732	737D17	JT8D-15
B733	B733	7373B2	CFM56-3B1
B734	B734	737400	CFM56-3C1
B735	B735	737500	CFM56-3C1
B736	B736	737500	CFM56-7B20
B737	B737	737700	CFM56-7B24
B738	B738	737700	CFM56-7B26
B739	B738	737700	CFM56-7B26
B73A	B732	737D17	JT8D-15
B73B	B732	737D17	JT8D-15
B73C	B732	737D17	JT8D-15
B73D	B732	737D17	JT8D-15
B73E	B732	737D17	JT8D-15
B73F	B732	737D17	JT8D-15
B73G	B732	737D17	JT8D-15
B73H	B732	737D17	JT8D-15
B73J	B732	737D17	JT8D-15
B73K	B732	737D17	JT8D-15
B73M	B732	737D17	JT8D-15
B73O	B732	737D17	JT8D-15
B73Q	B732	737D17	JT8D-15
B73R	B732	737D17	JT8D-15
B73S	B732	737D17	JT8D-15
B73T	B732	737D17	JT8D-15
B73U	B732	737D17	JT8D-15
B73V	B732	737D17	JT8D-15
B73W	B732	737D17	JT8D-15
B740	B742	74710Q	JT9D-7A
B741	B742	74710Q	JT9D-7A
B742	B742	747200	CF6-50E2
B743	B743	74720A	JT9D-7R4G2
B744	B744	747400	CF6-80C2B1F
B745	B744	747400	CF6-80C2B1F
B746	B744	747400	CF6-80C2B1F
B747	B744	747400	CF6-80C2B1F
B748	B744	747400	CF6-80C2B1F
B749	B744	747400	CF6-80C2B1F
B74A	B742	747200	CF6-50E2
B74B	B742	747200	CF6-50E2
B74D	B742	747200	CF6-50E2
B74E	B742	747200	CF6-50E2
B74F	B742	747200	CF6-50E2
B74Q	B742	747200	CF6-50E2
B74R	B744	747SP	CF6-45A2
B74S	B744	747SP	JT9D-7A
B74W	B744	747SP	CF6-45A2
B750	B752	757RR	RB211-535E4
B751	B752	757RR	RB211-535E4
B752	B752	757RR	RB211-535E4
B753	B753	757RR	RB211-535E4-C
B754	B753	757RR	RB211-535E4-C
B755	B753	757RR	RB211-535E4-C

B756	B753	757RR	RB211-535E4-C
B757	B753	757RR	RB211-535E4-C
B758	B753	757RR	RB211-535E4-C
B759	B753	757RR	RB211-535E4-C
B75B	B752	757RR	RB211-535E4
B75E	B752	757RR	RB211-535E4
B75F	B752	757RR	RB211-535E4
B75L	B752	757RR	RB211-535E4
B75Q	B752	757RR	RB211-535E4
B75R	B752	757RR	RB211-535E4
B75S	B752	757RR	RB211-535E4
B75T	B752	757RR	RB211-535E4
B75V	B752	757RR	RB211-535E4
B760	B762	767CF6	CF6-80A
B761	B762	767CF6	CF6-80A
B762	B762	767CF6	CF6-80A
B763	B763	767CF6	PW4060
B764	B763	767CF6	PW4060
B765	B763	767CF6	PW4060
B766	B763	767CF6	PW4060
B767	B763	767CF6	PW4060
B768	B763	767CF6	PW4060
B769	B763	767CF6	PW4060
B76E	B763	767CF6	PW4060
B76L	B763	767CF6	PW4060
B76Q	B763	767CF6	PW4060
B76R	B763	767CF6	PW4060
B76S	B763	767CF6	PW4060
B770	B772	777200	GE90-90B
B771	B772	777200	GE90-90B
B772	B772	777200	GE90-90B
B773	B773	777300	TRENT892
B774	B773	777300	TRENT892
B775	B773	777300	TRENT892
B776	B773	777300	TRENT892
B777	B773	777300	TRENT892
B778	B773	777300	TRENT892
B77B	B772	777200	GE90-90B
B77E	B772	777200	GE90-90B
B77H	B772	777200	GE90-90B
B77Q	B772	777200	GE90-90B
B77W	B772	777200	GE90-90B
BA10	BA11	BAC111	SPEYMK512DW
BA11	BA11	BAC111	SPEYMK512DW
BA12	BA11	BAC111	SPEYMK512DW
BA13	BA11	BAC111	SPEYMK512DW
BA14	BA11	BAC111	SPEYMK512DW
BA45	B462	BAE146	ALF502R-5
BA46	B462	BAE146	ALF502R-5
BE08	PAY2	DHC6	PT6A-20
BE09	PAY2	DHC6	PT6A-20
BE10	PAY2	DHC6	PT6A-20
BE11	PAY2	DHC6	PT6A-20
BE18	PA31	BEC58P	IO-540-K1B5
BE1O	PAY2	DHC6	PT6A-20
BE20	BE20	DHC6	PT6A-42
BE2O	BE20	DHC6	PT6A-42
BE30	SH36	SD330	PT6A-60A
BE31	P28A	CNA172	IO-360-B
BE32	P28A	CNA172	IO-360-B
BE33	P28A	CNA172	IO-360-B
BE34	P28A	CNA172	IO-360-B
BE35	P28A	CNA172	IO-360-B
BE36	P28A	CNA172	IO-360-B
BE38	P28A	CNA172	IO-360-B
BE39	P28A	CNA172	IO-360-B
BE3O	SH36	SD330	PT6A-60A
BE40	C560	CNA55B	JT15D-5
BE55	PA31	BEC58P	IO-540-K1B5
BE56	PA31	BEC58P	IO-540-K1B5
BE76	PA34	BEC58P	IO-360-B
BE9	BE99	DHC6	PT6A-20
BE90	BE99	DHC6	PT6A-20
BE91	BE99	DHC6	PT6A-20
BE92	BE99	DHC6	PT6A-20
BE93	BE99	DHC6	PT6A-20
BE94	BE99	DHC6	PT6A-20
BE95	BE99	DHC6	PT6A-20
BE96	BE99	DHC6	PT6A-20
BE97	BE99	DHC6	PT6A-20
BE98	BE99	DHC6	PT6A-20
BE99	BE99	DHC6	PT6A-20
BE9L	BE9L	DHC6	PT6A-20
BE9T	BE9L	DHC6	PT6A-20
BEL9	BE9L	DHC6	PT6A-20
BELF	C130	C130	PT6A-20
BET	BE99	DHC6	PT6A-20
BN2P	PA31	BEC58P	O-540-E4C5
BN2T	PAY2	DHC6	ASN-250-B17C
C130	C130	C130	ASN-501-D22A

C172	P28A	CNA172	IO-360-B
C182	P28A	CNA172	IO-540-K1B5
C206	TRIN	CNA206	IO-540-K1B5
C208	D228	GASEPF	PT6A-27
C210	P28A	CNA172	IO-540-K1B5
C212	PAY2	DHC6	TPE331-10-511C
C21O	P28A	CNA172	IO-540-K1B5
C2O6	TRIN	CNA206	IO-540-K1B5
C2O8	D228	GASEPF	PT6A-27
C310	PA27	BEC58P	IO-540-K1B5
C31O	PA27	BEC58P	IO-540-K1B5
C335	PA31	BEC58P	IO-540-K1B5
C340	PA31	BEC58P	IO-540-K1B5
C34O	PA31	BEC58P	IO-540-K1B5
C401	PA31	BEC58P	IO-540-K1B5
C402	PA31	BEC58P	IO-540-K1B5
C404	PA31	BEC58P	IO-540-K1B5
C411	PA31	BEC58P	IO-540-K1B5
C414	PA31	BEC58P	IO-540-K1B5
C421	C421	BEC58P	IO-540-K1B5
C425	PAY2	DHC6	PT6A-27
C441	PAY2	DHC6	PT6A-27
C46	AT72	HS748A	R-2800-75
C4O2	PA31	BEC58P	IO-540-K1B5
C500	C550	CNA500	JT15D-4B
C501	C550	CNA500	JT15D-4B
C502	C550	CNA500	JT15D-4B
C503	C550	CNA500	JT15D-4B
C505	C550	CNA500	JT15D-4B
C506	C550	CNA500	JT15D-4B
C508	C550	CNA500	JT15D-4B
C525	C550	CNA500	FJ44-1A
C550	C550	CNA55B	JT15D-4
C551	C550	CNA55B	JT15D-4B
C552	C550	CNA55B	JT15D-4B
C553	C550	CNA55B	JT15D-4B
C555	C550	CNA55B	JT15D-4B
C556	C550	CNA55B	JT15D-4B
C559	C550	CNA55B	JT15D-4B
C55O	C550	CNA55B	JT15D-4B
C560	C560	CNA55B	JT15D-5D
C561	C560	CNA55B	JT15D-5D
C563	C560	CNA55B	JT15D-5D
C565	C560	CNA55B	JT15D-5D
C566	C560	CNA55B	JT15D-5D
C569	C560	CNA55B	JT15D-5D
C56B	C560	CNA55B	JT15D-5D
C56C	C560	CNA55B	JT15D-5D
C56E	C560	CNA55B	JT15D-5D
C56G	C560	CNA55B	JT15D-5D
C56K	C560	CNA55B	JT15D-5D
C56L	C560	CNA55B	JT15D-5D
C56O	C560	CNA55B	JT15D-5D
C56Q	C560	CNA55B	JT15D-5D
C56S	C560	CNA55B	JT15D-5D
C56X	C560	CNA55B	JT15D-5D
C650	C560	CNA55B	TFE731-3B-100S
C750	C750	CNA750	AE-3007C
C75O	C750	CNA750	AE-3007C
C97	AT72	DC6	R-4360
CARJ	BA11	EMB145	CF34-3B1
CARV	AT72	DC6	R-2000-11
CCJ	CL60	CL601	CF34-3B1
CJ1	C550	CNA500	CF34-3B1
CJ2	C550	CNA500	CF34-3B1
CJR1	CRJ1	CL601	CF34-3A1
CJR2	CRJ1	CL601	CF34-3B1
CJR7	CRJ1	CL601	CF34-8C5
CJRJ	BA11	EMB145	CF34-3B1
CL44	C130	C130	TYNE-515\10
CL60	CL60	CL601	CF34-3B1
CL61	CL60	CL601	CF34-3B1
CL62	CL60	CL601	CF34-3B1
CL63	CL60	CL601	CF34-3B1
CL64	CL60	CL601	CF34-3B1
CL65	CL60	CL601	CF34-3B1
CL66	CL60	CL601	CF34-3B1
CL69	CL60	CL601	CF34-3B1
CLVT	SB20	CVR580	ASN501-D13H
CN35	SF34	SF340	CT7-7A
CONC	B752	CONCRD	OLYMPUS593
CONI	AT72	DC6	R-3350-91
CR1	CRJ1	CL601	CF34-3B1
CR2	CRJ1	CL601	CF34-3B1
CR2J	CRJ1	CL601	CF34-3B1
CR3	CRJ1	CL601	CF34-3B1
CR4	CRJ1	CL601	CF34-8C1
CR6	CRJ1	CL601	CF34-8C1
CR7	CRJ1	CL601	CF34-8C1
CR7J	CRJ1	CL601	CF34-8C1

CR9	CRJ1	CL601	CF34-8C1
CRAJ	BA11	EMB145	CF34-3B1
CRG1	CRJ1	CL601	CF34-3B1
CRG2	CRJ1	CL601	CF34-3B1
CRG7	CRJ1	CL601	CF34-8C1
CRJ	CRJ1	CL601	CF34-3A1
CRJ1	CRJ1	CL601	CF34-3A1
CRJ2	CRJ1	CL601	CF34-3B1
CRJ3	CRJ1	CL601	CF34-3B1
CRJ4	CRJ1	CL601	CF34-3B1
CRJ5	CRJ1	CL601	CF34-8C5
CRJ6	CRJ1	CL601	CF34-8C5
CRJ7	CRJ1	CL601	CF34-8C5
CRJ8	CRJ1	CL601	CF34-8C5
CRJ9	CRJ1	CL601	CF34-8C5
CRJI	CRJ1	CL601	CF34-3B1
CRRJ	BA11	EMB145	CF34-3B1
CS5	SF34	SF340	CT7-9C
CULT	SB20	CVR580	ASN501-D13H
CV99	B703	707320	CJ-805-3
CVIT	SB20	CVR580	ASN501-D13H
CVL	SB20	CVR580	ASN501-D13H
CVLP	AT43	DC6	R-2800-CB17
CVLT	SB20	CVR580	ASN501-D13H
CVRT	SB20	CVR580	ASN501-D13H
CVT	SB20	CVR580	ASN501-D13H
CVTL	SB20	CVR580	ASN501-D13H
CZLT	SB20	CVR580	ASN501-D13H
CZRJ	BA11	EMB145	CF34-3B1
D028	D228	DHC6	TPE331-5-252D
D032	D328	DHC8	PW119B
D228	D228	DHC6	TPE331-5-252D
D328	D328	DHC8	PW119B
D8A	DH8A	DHC8	PW120A
D8B	DH8A	DHC8	PW123
D8C	DH8C	DHC830	PW123
D8H	DH8A	DHC8	PW120A
D8HB	DH8A	DHC8	PW120A
D8Q	DH8C	DHC830	PW123
DC03	DH8C	DC3	R-2000-11
DC08	DC87	DC850	JT3D-3B
DC10	DC10	DC1010	CF6-50C2
DC3	DH8C	DC3	R-2000-11
DC4	AT72	DC6	R-2000-11
DC6	AT72	DC6	R-2800-CB17
DC7	AT72	DC6	R-3350-18EA1
DC8	DC87	DC850	JT3D-3B
DC80	DC87	DC850	JT3D-3B
DC81	DC87	DC850	JT3D-3B
DC82	DC87	DC850	JT3D-3B
DC83	DC87	DC850	JT3D-3B
DC84	DC87	DC850	JT3D-3B
DC85	DC87	DC850	JT3D-3B
DC86	DC87	DC850	JT3D-3B
DC87	DC87	DC870	CFM56-2C1
DC88	DC87	DC870	CFM56-2C1
DC8A	DC87	DC850	JT3D-3B
DC8B	DC87	DC850	JT3D-3B
DC8C	DC87	DC850	JT3D-3B
DC8G	DC87	DC850	JT3D-3B
DC8H	DC87	DC850	JT3D-3B
DC8L	DC87	DC850	JT3D-3B
DC8Q	DC87	DC850	JT3D-3B
DC8S	DC87	DC850	JT3D-3B
DC9	DC94	DC910	JT8D-7B
DC90	DC94	DC910	JT8D-7B
DC91	DC94	DC910	JT8D-7B
DC92	DC94	DC910	JT8D-11
DC93	DC94	DC910	JT8D-9A
DC94	DC94	DC910	JT8D-11
DC95	DC94	DC910	JT8D-17
DC96	DC94	DC910	JT8D-17
DC97	DC94	DC910	JT8D-17
DC98	DC94	DC910	JT8D-217C
DC99	DC94	DC910	JT8D-17
DC9A	DC94	DC910	JT8D-17
DC9B	DC94	DC910	JT8D-17
DC9C	DC94	DC910	JT8D-17
DC9D	DC94	DC910	JT8D-17
DC9G	DC94	DC910	JT8D-17
DC9I	DC94	DC910	JT8D-17
DC9M	DC94	DC910	JT8D-17
DC9O	DC94	DC910	JT8D-17
DC9Q	DC94	DC910	JT8D-17
DC9S	DC94	DC910	JT8D-17
DC9W	DC94	DC910	JT8D-17
DCH6	PAY2	DHC6	PT6A-27
DCH7	AT72	DHC7	PT6A-50
DCH8	DH8A	DHC8	PW120A
DH06	PAY2	DHC6	PT6A-27

DH08	DH8A	DHC8	PW120A
DH6	PAY2	DHC6	PT6A-27
DH6A	PAY2	DHC6	PT6A-27
DH6C	PAY2	DHC6	PT6A-27
DH6L	PAY2	DHC6	PT6A-27
DH7	AT72	DHC7	PT6A-50
DH7A	AT72	DHC7	PT6A-50
DH7C	AT72	DHC7	PT6A-50
DH7D	AT72	DHC7	PT6A-50
DH7M	AT72	DHC7	PT6A-50
DH8	DH8A	DHC8	PW120A
DH8A	DH8A	DHC8	PW120A
DH8B	DH8A	DHC8	PW123C
DH8C	DH8C	DHC830	PW123
DH8D	DH8C	DHC830	PW150A
DH8H	DH8C	DHC830	PW150A
DH8I	DH8C	DHC830	PW150A
DH8L	DH8C	DHC830	PW150A
DH8M	DH8C	DHC830	PW150A
DH8Q	DH8C	DHC830	PW150A
DH8R	DH8C	DHC830	PW150A
DH8S	DH8C	DHC830	PW150A
DH8T	DH8C	DHC830	PW150A
DH8X	DH8C	DHC830	PW150A
DH8Z	DH8C	DHC830	PW150A
DHA8	DH8A	DHC8	PW120A
DHB8	DH8A	DHC8	PW123C
DHC2	BE9L	GASEPV	PT6A-20
DHC3	D228	GASEPF	PT6A-20
DHC4	DH8C	DC3	R-2000-11
DHC6	PAY2	DHC6	PT6A-27
DHC7	AT72	DHC7	PT6A-50
DHC8	DH8C	DHC8	PW120A
DHG8	DH8C	DHC830	PW150A
DHS8	DH8C	DHC830	PW150A
DHV8	DH8C	DHC830	PW150A
DO28	D228	DHC6	TPE331-5-252D
DO32	D328	DHC8	PW119B
DSH6	PAY2	DHC6	PT6A-27
DSH8	DH8A	DHC8	PW120A
E110	PAY2	DHC6	PT6A-34
E11O	PAY2	DHC6	PT6A-34
E120	E120	EMB120	PW118
E135	E145	EMB145	AE3007A1\2
E145	E145	EMB145	AE3007A1\2
E170	E145	EMB145	AE3007A1\2
ERD	CRJ1	EMB145	AE3007A3
F100	F100	F10062	TAYMK 650-15
F27	F27	HS748A	PW125B
F27A	F27	HS748A	PW125B
F27H	F27	HS748A	PW125B
F27T	F27	HS748A	PW125B
F28	F28	F28MK2	TAYMK 650-15
F28A	F28	F28MK2	TAYMK 650-15
F28C	F28	F28MK2	TAYMK 650-15
F28F	F28	F28MK2	TAYMK 650-15
F2TH	E145	EMB145	CFE738-1-1B
F50	F50	DHC830	PW125B
F70	F70	F10062	TAYMK 620-15
F90	PAY2	DHC6	PT6A-20
F900	F900	EMB145	TFE731-5BR-1C
F90B	PAY2	DHC6	PT6A-20
FA10	FA10	LEAR35	TFE731-2-1C
FA20	FA20	FAL20	CF700-2D
FA50	FA20	FAL20	CF700-2D
FK27	F27	HS748A	PW125B
FK28	F28	F28MK2	TAYMK 650-15
FK70	F70	F10062	TAYMK 620-15
FT2H	E145	EMB145	CFE738-1-1B
FTH2	E145	EMB145	CFE738-1-1B
G159	AT72	HS748A	DART529-8E
G21	PAY2	DHC6	PT6A-27
G21A	PAY2	DHC6	PT6A-27
G73	PA31	DHC6	PT6A-27
G73T	PAY2	DHC6	PT6A-27
GFL2	F70	GII	SPEYMK 511-8
GFL3	F70	GIIB	SPEYMK 511-8
GFL4	F70	GIV	TAY-MK -611-8
GFL5	F70	GV	BR710A1-10
GLF2	F70	GII	SPEYMK 511-8
GLF3	F70	GIIB	SPEYMK 511-8
GLF4	F70	GIV	TAY-MK -611-8
GLF5	F70	GV	BR710A1-10
GU2	F70	GII	SPEYMK 511-8
GU3	F70	GIIB	SPEYMK 511-8
GUL2	F70	GII	SPEYMK 511-8
GUL3	F70	GIIB	SPEYMK 511-8
GUL4	F70	GIV	TAY-MK -611-8
GUL5	F70	GV	BR710A1-10
H25	H25B	CNA55B	TFE731-5BR

H25A	H25B	CNA55B	TFE731-5BR
H25B	H25B	CNA55B	TFE731-3-1H
H25C	H25B	CNA55B	TFE731-5BR
H25D	H25B	CNA55B	TFE731-5BR
H25E	H25B	CNA55B	TFE731-5BR
H25G	H25B	CNA55B	TFE731-5BR
H25I	H25B	CNA55B	TFE731-5BR
H25J	H25B	CNA55B	TFE731-5BR
H25L	H25B	CNA55B	TFE731-5BR
H25M	H25B	CNA55B	TFE731-5BR
H25N	H25B	CNA55B	TFE731-5BR
H25P	H25B	CNA55B	TFE731-5BR
H25R	H25B	CNA55B	TFE731-5BR
H25S	H25B	CNA55B	TFE731-5BR
H25T	H25B	CNA55B	TFE731-5BR
H25U	H25B	CNA55B	TFE731-5BR
H25V	H25B	CNA55B	TFE731-5BR
H25X	H25B	CNA55B	TFE731-5BR
HERN	PAY2	DHC6	G Q -30-MK 2
HF20	C560	CNA55B	CJ610-1
HZ5B	H25B	CNA55B	TFE731-3-1H
I114	AT72	HS748A	TV7-117C
Il62	B703	707QN	NK-8-4
IL18	C130	C130	AI-20
IL62	B703	707QN	NK-8-4
IL76	A30B	DC8QN	D-30KP
IL86	A343	A340	NK-86
IL96	A343	A340	PS-90A
J31	JS31	DHC6	TPE331-12UA-701
J32	JS31	SD330	TPE331-12UA-701
J328	D328	DHC8	PW119B
JS20	JS31	SD330	ASTAZOUXVI D
JS31	JS31	SD330	TPE331-10UG-513
JS32	JS31	SD330	TPE331-12UA-701
JS41	JS41	SF340	TPE331-14GR
JST	JS31	DHC6	TPE331-10UG-513
L101	L101	L1011	RB211-22B
L10A	L101	L1011	RB211-22B
L188	C130	L188	ASN501-D13A
L1O1	L101	L1011	RB211-22B
L410	PAY2	DHC6	M-601E
L4T	PAY2	DHC6	M-601E
L610	AT43	DHC8	M-602
LJ23	LJ35	LEAR25	CJ610-4
LJ24	LJ35	LEAR25	CJ610-6
LJ25	LJ35	LEAR25	CJ610-6
LJ28	LJ35	LEAR25	CJ610-8A
LJ29	LJ35	LEAR25	CJ610-8A
LJ30	LJ35	LEAR35	TFE731-2-3B
LJ31	LJ35	LEAR35	TFE731-2-3B
LJ32	LJ35	LEAR35	TFE731-2-3B
LJ33	LJ35	LEAR35	TFE731-2-3B
LJ34	LJ35	LEAR35	TFE731-2-3B
LJ35	LJ35	LEAR35	TFE731-2-3B
LJ36	LJ35	LEAR35	TFE731-2-3B
LJ37	LJ35	LEAR35	TFE731-2-3B
LJ38	LJ35	LEAR35	TFE731-2-3B
LJ39	LJ35	LEAR35	TFE731-2-3B
LJ45	LJ45	LEAR35	TFE731-20
LJ55	LJ35	LEAR35	TFE731-3A-2B
LJ60	LJ45	LEAR35	PW305A
LR24	LJ35	LEAR25	CJ610-6
LR25	LJ35	LEAR25	CJ610-6
LR35	LJ35	LEAR35	TFE731-2-3B
LR45	LJ45	LEAR35	TFE731-20
M20	P28A	CNA172	IO-360-B
M20A	P28A	CNA172	IO-360-B
M20B	P28A	CNA172	IO-360-B
M20C	P28A	CNA172	IO-360-B
M20D	P28A	CNA172	IO-360-B
M20E	P28A	CNA172	IO-360-B
M20F	P28A	CNA172	IO-360-B
M20G	P28A	CNA172	IO-360-B
M20I	P28A	CNA172	IO-540-K1B5
M20J	P28A	CNA172	IO-540-K1B5
M20K	P28A	CNA172	IO-540-K1B5
M20L	P28A	CNA172	IO-540-K1B5
M20M	P28A	CNA172	IO-540-K1B5
M20N	P28A	CNA172	IO-540-K1B5
M20P	P28A	CNA172	IO-540-K1B5
M20R	P28A	CNA172	IO-540-K1B5
M20S	P28A	CNA172	IO-540-K1B5
M20T	P28A	CNA172	IO-540-K1B5
M20U	P28A	CNA172	IO-360-B
M28	SH36	SD330	TWD-10B
M2OE	P28A	CNA172	IO-360-B
M2OK	P28A	CNA172	IO-540-K1B5
M2OP	P28A	CNA172	IO-540-K1B5
M2OT	P28A	CNA172	IO-540-K1B5
M404	AT72	HS748A	R-2800-CB16

MD11	MD11	MD11GE	CF6-80C2D1F
MD80	MD82	MD81	JT8D-217C
MD81	MD82	MD81	JT8D-217C
MD82	MD82	MD82	JT8D-217C
MD83	MD83	MD83	JT8D-219
MD84	MD83	MD83	JT8D-219
MD85	MD83	MD83	JT8D-219
MD86	MD83	MD83	JT8D-219
MD87	MD83	MD83	JT8D-217C
MD88	MD83	MD83	JT8D-219
MD89	MD83	MD83	JT8D-219
MD8I	MD82	MD82	JT8D-219
MD8O	MD82	MD81	JT8D-217C
MD90	MD83	MD9025	V2525-D5
MD91	MD83	MD9025	V2525-D5
MD92	MD83	MD9025	V2525-D5
MD93	MD83	MD9025	V2525-D5
MDII	MD11	MD11GE	CF6-80C2D1F
MU02	MU2	DHC6	TPE331-10-501C
MU2	MU2	DHC6	TPE331-10-501C
MU2A	MU2	DHC6	TPE331-10-501C
MU2B	MU2	DHC6	TPE331-10-501C
MU2G	MU2	DHC6	TPE331-10-501C
MU2I	MU2	DHC6	TPE331-10-501C
MU2N	MU2	DHC6	TPE331-10-501C
MU2R	MU2	DHC6	TPE331-10-501C
MU2T	MU2	DHC6	TPE331-10-501C
MU3	C550	MU3001	JT15D-4D
MU30	C550	MU3001	JT15D-4D
N22	MU2	DHC6	ASN250-B17B
N24	MU2	DHC6	ASN250-B17B
N24A	MU2	DHC6	ASN250-B17B
N260	SH36	SD330	BASTANVIC
N262	SH36	SD330	BASTANVIC
N264	SH36	SD330	BASTANVIC
N265	SH36	SD330	BASTANVIC
NOMA	D228	DHC6	ASN250-B17B
P20	P28A	CNA172	IO-320-DIAD
P20R	P28A	CNA172	IO-320-DIAD
P23	PA27	BEC58P	IO-540-K1B5
P23A	PA27	BEC58P	IO-540-K1B5
P23R	PA27	BEC58P	IO-540-K1B5
P24	TRIN	CNA206	IO-540-K1B5
P24A	TRIN	CNA206	IO-540-K1B5
P24T	TRIN	CNA206	IO-540-K1B5
P27	PA27	BEC58P	IO-540-K1B5
P27A	PA27	BEC58P	IO-540-K1B5
P27R	PA27	BEC58P	IO-540-K1B5
P28	P28A	CNA172	IO-360-B
P28A	P28A	CNA172	IO-360-B
P28B	P28A	CNA172	IO-360-B
P28C	P28A	CNA172	IO-360-B
P28P	P28A	CNA172	IO-360-B
P28R	P28A	CNA172	IO-360-B
P28T	P28A	CNA172	IO-360-B
P30	P28A	CNA172	IO-320-DIAD
P30A	P28A	CNA172	IO-320-DIAD
P30B	P28A	CNA172	IO-320-DIAD
P30T	P28A	CNA172	IO-320-DIAD
P31	PA31	BEC58P	IO-540-K1B5
P31A	PA31	BEC58P	IO-540-K1B5
P31C	PA31	BEC58P	IO-540-K1B5
P31P	PA31	BEC58P	IO-540-K1B5
P31R	PA31	BEC58P	IO-540-K1B5
P31T	PAY2	DHC6	PT6A-28
P32	P28A	CNA172	IO-540-K1C5
P32A	P28A	CNA172	IO-540-K1C5
P32B	P28A	CNA172	IO-540-K1C5
P32R	P28A	CNA172	IO-540-K1C5
P32T	P28A	CNA172	IO-540-K1C5
P34	PA34	BEC58P	IO-360-B
P34A	PA34	BEC58P	IO-360-B
P34R	PA34	BEC58P	IO-360-B
P34T	PA34	BEC58P	IO-360-B
P44	PA34	CNA206	IO-360-B
P44A	PA34	CNA206	IO-360-B
P44R	PA34	CNA206	IO-360-B
P46	TRIN	CNA206	IO-540-K1B5
P46A	TRIN	CNA206	IO-540-K1B5
P46D	TRIN	CNA206	IO-540-K1B5
P46P	TRIN	CNA206	IO-540-K1B5
P46R	TRIN	CNA206	IO-540-K1B5
P46T	TRIN	CNA206	PT6A-42
P68	PA34	CNA206	IO-360-B
P68A	PA34	CNA206	IO-360-B
P68B	PA34	CNA206	IO-360-B
P68C	PA34	CNA206	IO-360-B
P68T	PA34	CNA206	IO-360-B
P808	C560	CNA55B	VIPER-MK-526
PA20	P28A	CNA172	IO-320-DIAD

PA22	P28A	CNA172	IO-320-DIAD
PA23	PA27	BEC58P	IO-540-K1B5
PA24	TRIN	CNA206	IO-540-K1B5
PA27	PA27	BEC58P	IO-540-K1B5
PA28	P28A	CNA172	IO-360-B
PA30	P28A	CNA172	IO-320-DIAD
PA31	PA31	BEC58P	IO-540-K1B5
PA32	P28A	CNA172	IO-540-K1C5
PA34	PA34	BEC58P	IO-360-B
PA38	P28A	CNA172	IO-320-DIAD
PA44	PA34	CNA206	IO-360-B
PA46	TRIN	CNA206	IO-540-K1B5
PAT	PAY2	DHC6	PT6A-28
PAY1	PAY2	DHC6	PT6A-28
PAY2	PAY2	DHC6	PT6A-28
PAY3	PAY3	DHC6	PT6A-41
PC12	D228	DHC6	PT6A-67D
PC6	BE9L	CNA20T	PT6A-27
PC6T	BE9L	CNA20T	PT6A-27
PRM1	C550	CNA500	FJ44-2A
RJ1	B462	BAE146	LF507-1H
RJ1H	B462	BAE146	LF507-1H
RJ70	B462	BAE146	LF507-1H
RJ85	B462	BAE146	LF507-1H
S210	B732	737	JT8D-7
S601	C550	CNA500	JT15D-4
SA26	PAY2	DHC6	TPE331-3U-303G
SA27	PAY2	DHC6	TPE331-11U-611G
SABR	C560	CNA55B	JT12A-8
SBR	C560	CNA55B	JT12A-8
SBR1	C560	CNA55B	JT12A-8
SBR2	C560	CNA55B	JT12A-8
SBR3	C560	CNA55B	JT12A-8
SBR4	C560	CNA55B	JT12A-8
SBR6	C560	CNA55B	JT12A-8
SBR9	C560	CNA55B	JT12A-8
SBRE	C560	CNA55B	JT12A-8
SBRI	C560	CNA55B	JT12A-8
SBRL	C560	CNA55B	JT12A-8
SC7	PAY2	DHC6	TPE331-2-201A
SC7A	PAY2	DHC6	TPE331-2-201A
SD30	SH36	SD330	PT6A-45R
SD33	SH36	SD330	PT6A-45R
SD34	SH36	SD330	PT6A-45R
SD35	SH36	SD330	PT6A-45R
SD36	SH36	SD330	PT6A-45R
SD38	SH36	SD330	PT6A-45R
SF20	SB20	CVR580	AE2100A
SF34	SF34	SF340	CT7-9B
SGUP	C130	C130	ASN-501-D22C
SH30	SH36	SD330	PT6A-45R
SH31	SH36	SD330	PT6A-45R
SH32	SH36	SD330	PT6A-45R
SH33	SH36	SD330	PT6A-45R
SH34	SH36	SD330	PT6A-45R
SH35	SH36	SD330	PT6A-45R
SH36	SH36	SD330	PT6A-67R
SH37	SH36	SD330	PT6A-67R
SH38	SH36	SD330	PT6A-67R
SW3	SW3	DHC6	TPE331-10U-503G
SW3A	SW3	DHC6	TPE331-10U-503G
SW4	SW3	DHC6	TPE331-12U-701G
SW4A	SW3	DHC6	TPE331-12U-701G
T134	T134	F10065	D-30-3
T154	T154	727D17	D-30KU-154-II
T20	B752	757RR	PS-90AT
T204	B752	757RR	RB211-535E4-B
T34P	BE9L	GASEPV	PT6A-20
T34T	BE9L	GASEPV	PT6A-20
TU34	T134	F10065	D-30-3
TU5	T154	727D17	D-30KU-154-II
TU54	T154	727D17	D-30KU-154-II
VC10	DC87	DC870	CONWAY42MK540
VF14	E145	EMB145	M45HMK 501
VISC	AT72	HS748A	DART6MK510
WW24	H25B	IA1125	TFE731-3-1G
WW25	H25B	IA1125	TFE731-3-1G
WW26	H25B	IA1125	TFE731-3-1G
Y12	PAY2	DHC6	PT6A-27
Y40	B722	727100	AI-25
Y42	B722	727100	D-36
Y42K	B722	727100	D-36
Y8	AT72	DHC7	AI-20M
YA40	B722	727100	AI-25
YA42	B722	727100	D-36
YK4	B722	727100	AI-25
YK40	B722	727100	AI-25
YK42	B722	727100	D-36
YN2	PAY2	DHC6	PT6A-27
YN7	AT72	HS748A	WJ5A-1

YS11 AT72 HS748A DART542-10J\K

APPENDIX E: Aircraft Code Mappings used for Engine Assignments

In order to develop distributions of engines for assignments to flights, BACK aircraft codes must be mapped to ETMS and OAG aircraft codes. The reason for using a unique listing of BACK aircraft codes to serve as the basis for this mapping is to make sure all of the BACK aircraft types have been accounted and no overlaps occur in the mappings. This helps to ensure proper development of distributions. The mappings range from exact matches to approximate substitutions.

BACK Aircraft Code	ETMS or OAG Aircraft Code
-27	F27
1121A-JET-COMM	AC11
1121B-JET-COMM	AC11
1121JETCOMM	AC11
1121-JET-COMM	AC11
328-300	A323
328-310	A323
390PREMIER	PRM1
390-PREMIER	PRM1
707-123B	B701
707-138B	B701
707-307C	B703
707-312B	B703
707-320C	B703
707-321	B703
707-321B	B703
707-321C	B703
707-323B	B703
707-323C	B703
707-324C	B703
707-327C	B703
707-328B	B703
707-328C	B703
707-329	B703
707-329C	B703
707-330B	B703
707-330C	B703
707-331B	B703
707-331C	B703
707-336C	B703
707-337C	B703
707-338C	B703
707-340C	B703
707-341C	B703
707-344C	B703
707-345C	B703
707-347C	B703
707-351B	B703
707-351C	B703
707-353B	B703
707-355C	B703
707-358C	B703
707-365C	B703
707-366C	B703
707-368C	B703
707-369C	B703
707-370C	B703
707-373C	B703
707-379C	B703
707-382B	B703
707-384C	B703
707-385C	B703
707-386C	B703
707-387B	B703
707-387C	B703
707-396C	B703
707-399C	B703
707-3B4C	B703
707-3F5C	B703
707-3H7C	B703
707-3J6B	B703
707-3J6C	B703
707-3J8C	B703
707-3J9C	B703
707-3K1C	B703
707-3L5C	B703
707-3L6B	B703

707-3L6C	B703
707-3M1C	B703
707-3P1C	B703
707-3W6C	B703
707-E3A	B703
707-E3C	B703
707-E3D	B703
707-E3F	B703
707-E6A	B703
707-E8A	B703
707-E8B	B703
707-E8C	B703
707-GE8C	B703
707-KE3A	B703
707-TE8A	B703
707-YE8B	B703
717-200	B712
717-22A	B712
717-22K	B712
717-231	B712
717-23S	B712
717-2BD	B712
717-2BL	B712
717-2CM	B712
717-2K9	B712
720-023B	B720
720-047B	B720
720-051B	B720
727-108C	B721
727-109	B721
727-109C	B721
727-113C	B721
727-116C	B721
727-116F	B721
727-121C	B721
727-123	B721
727-123F	B721
727-134	B721
727-134C	B721
727-14	B721
727-151C	B721
727-155C	B721
727-17	B721
727-171C	B721
727-172C	B721
727-173C	B721
727-180C	B721
727-185C	B721
727-191	B721
727-193	B721
727-1A0	B721
727-1A7C	B721
727-1H2	B721
727-208	B722
727-21	B722
727-212	B722
727-212F	B722
727-214	B722
727-214F	B722
727-217F	B722
727-21C	B722
727-21F	B722
727-22	B722
727-221	B722
727-221F	B722
727-222	B722
727-222F	B722
727-223	B722
727-223F	B722
727-224	B722
727-224F	B722
727-225	B722
727-225F	B722
727-227	B722
727-227F	B722
727-228	B722
727-228F	B722
727-22C	B722
727-22F	B722
727-23	B722
727-230	B722
727-230F	B722
727-231	B722
727-231F	B722
727-232	B722
727-232F	B722
727-233F	B722
727-235	B722
727-23F	B722
727-243	B722

727-243F	B722
727-247	B722
727-247F	B722
727-24C	B722
727-25	B722
727-251	B722
727-251F	B722
727-256	B722
727-259	B722
727-259F	B722
727-25C	B722
727-25F	B722
727-260	B722
727-260F	B722
727-264	B722
727-264F	B722
727-269	B722
727-27	B722
727-270	B722
727-276	B722
727-276F	B722
727-277	B722
727-277F	B722
727-27C	B722
727-27F	B722
727-281	B722
727-281F	B722
727-282	B722
727-282F	B722
727-286	B722
727-287	B722
727-287F	B722
727-290	B722
727-291F	B722
727-294	B722
727-29C	B722
727-29F	B722
727-2A1	B722
727-2A1F	B722
727-2A7F	B722
727-2B6	B722
727-2B6F	B722
727-2B7	B722
727-2D3	B722
727-2D3F	B722
727-2D4F	B722
727-2D6	B722
727-2F2	B722
727-2F2F	B722
727-2F9	B722
727-2F9F	B722
727-2H3	B722
727-2H3F	B722
727-2H9	B722
727-2J0	B722
727-2J0F	B722
727-2J4	B722
727-2J4F	B722
727-2J7	B722
727-2J7F	B722
727-2K3	B722
727-2K5	B722
727-2K5F	B722
727-2L4	B722
727-2L5	B722
727-2L8	B722
727-2M1	B722
727-2M7	B722
727-2M7F	B722
727-2N6	B722
727-2N8	B722
727-2P1	B722
727-2Q4F	B722
727-2Q6	B722
727-2Q6F	B722
727-2Q8	B722
727-2Q9F	B722
727-2R1	B722
727-2S2F	B722
727-2S7	B722
727-2T3	B722
727-2U5	B722
727-2X3F	B722
727-2X8	B722
727-2Y4	B722
727-30	B723
727-30C	B723
727-31	B723
727-31C	B723
727-31F	B723

727-35	B723
727-35F	B723
727-41	B724
727-41F	B724
727-44	B724
727-44C	B724
727-44F	B724
727-46	B724
727-46F	B724
727-51	B725
727-51C	B725
727-51F	B725
727-59F	B725
727-61	B722
727-62C	B726
727-64	B726
727-76	B727
727-76F	B727
727-77	B727
727-77C	B727
727-78	B727
727-78F	B727
727-81	B728
727-82	B728
727-82C	B728
727-82F	B728
727-86	B728
727-89	B728
727-90C	B729
727-95	B729
727-95F	B729
727-C3	B72C
737-112	B731
737-130	B731
737-201	B732
737-202C	B732
737-204	B732
737-204C	B732
737-205	B732
737-205C	B732
737-209	B732
737-210	B732
737-210C	B732
737-212	B732
737-214	B732
737-217	B732
737-219	B732
737-219C	B732
737-222	B732
737-228	B732
737-229	B732
737-229C	B732
737-230	B732
737-230C	B732
737-232	B732
737-236	B732
737-241	B732
737-242	B732
737-242C	B732
737-244	B732
737-244F	B732
737-247	B732
737-248	B732
737-248C	B732
737-258	B732
737-25A	B732
737-25C	B732
737-25CF	B732
737-260	B732
737-266	B732
737-268	B732
737-268C	B732
737-269	B732
737-270C	B732
737-275	B732
737-275C	B732
737-277	B732
737-27A	B732
737-281	B732
737-282	B732
737-282C	B732
737-284	B732
737-286	B732
737-286C	B732
737-287	B732
737-287C	B732
737-290C	B732
737-291	B732
737-293	B732
737-296	B732

737-297	B732
737-298C	B732
737-2A1	B732
737-2A1F	B732
737-2A3	B732
737-2A6	B732
737-2A8	B732
737-2A8C	B732
737-2A8F	B732
737-2A9C	B732
737-2B1	B732
737-2B1C	B732
737-2B2	B732
737-2B6	B732
737-2B6C	B732
737-2B7	B732
737-2C3	B732
737-2C9	B732
737-2D6	B732
737-2D6C	B732
737-2E1	B732
737-2E1F	B732
737-2E3	B732
737-2E7	B732
737-2F9	B732
737-2H3	B732
737-2H3C	B732
737-2H4	B732
737-2H4C	B732
737-2H5	B732
737-2H6	B732
737-2H6C	B732
737-2H7C	B732
737-2J8	B732
737-2J8C	B732
737-2K2	B732
737-2K2C	B732
737-2K3	B732
737-2K5	B732
737-2K6	B732
737-2K9	B732
737-2K9F	B732
737-2L7	B732
737-2L7C	B732
737-2L9	B732
737-2M2	B732
737-2M2C	B732
737-2M6	B732
737-2M6C	B732
737-2M8	B732
737-2M9	B732
737-2N0	B732
737-2N1	B732
737-2N3	B732
737-2N7	B732
737-2N8	B732
737-2N9C	B732
737-2P5	B732
737-2P6	B732
737-2Q2C	B732
737-2Q3	B732
737-2Q5C	B732
737-2Q8	B732
737-2Q8C	B732
737-2Q9	B732
737-2R4C	B732
737-2R6C	B732
737-2R8C	B732
737-2S2C	B732
737-2S3	B732
737-2S5C	B732
737-2S9	B732
737-2T2	B732
737-2T2C	B732
737-2T4	B732
737-2T4C	B732
737-2T5	B732
737-2T7	B732
737-2U4	B732
737-2U9	B732
737-2V5	B732
737-2V6	B732
737-2W8	B732
737-2X2	B732
737-2X6C	B732
737-2X9	B732
737-2Y5	B732
737-2Z6	B732
737-301	B733
737-306	B733

737-317	B733
737-319	B733
737-31B	B733
737-31L	B733
737-31S	B733
737-322	B733
737-329	B733
737-32Q	B733
737-330	N733
737-330F	N733
737-332	B733
737-33A	B733
737-33A(BW)	B733
737-33AF	B733
737-33R	B733
737-33S	B733
737-33V	B733
737-340	B733
737-341	B733
737-341F	B733
737-347	B733
737-348F	B733
737-34N	B733
737-34S	B733
737-35B	B733
737-35N	B733
737-36E	B733
737-36M	B733
737-36N	B733
737-36Q	B733
737-36R	B733
737-375	B733
737-376	B733
737-377	B733
737-377F	B733
737-37K	B733
737-37Q	B733
737-382	B733
737-382F	B733
737-38BF	B733
737-38J	B733
737-39A	B733
737-39K	B733
737-39M	B733
737-39MF	B733
737-39P	B733
737-3A1	B733
737-3A4	B733
737-3B3F	B733
737-3B7	B733
737-3G7	B733
737-3H4	B733
737-3H6F	B733
737-3H9	B733
737-3J6	B733
737-3K2	B733
737-3K9	B733
737-3L9	B733
737-3M8	B733
737-3M8F	B733
737-3Q4	B733
737-3Q4F	B733
737-3Q8	B733
737-3Q8F	B733
737-3S1	B733
737-3S3	B733
737-3S3F	B733
737-3T0	B733
737-3T0F	B733
737-3T5	B733
737-3U3	B733
737-3U8	B733
737-3W0	B733
737-3Y0	B733
737-3Y0F	B733
737-3Y5	B733
737-3Y9	B733
737-3Z0	B733
737-3Z8	B733
737-3Z9	B733
737-401	B734
737-405	B734
737-406	B734
737-408	B734
737-429	B734
737-42C	B734
737-42J	B734
737-42R	B734
737-430	B734
737-436	B734

737-43Q	B734
737-446	B734
737-448	B734
737-44P	B734
737-45D	B734
737-45R	B734
737-45S	B734
737-46B	B734
737-46J	B734
737-46M	B734
737-46N	B734
737-46Q	B734
737-476	B734
737-484	B734
737-48E	B734
737-490	B734
737-497	B734
737-49R	B734
737-4B3	B734
737-4B6	B734
737-4B7	B734
737-4C9	B734
737-4D7	B734
737-4H6	B734
737-4K5	B734
737-4L7	B734
737-4M0	B734
737-4Q3	B734
737-4Q8	B734
737-4S3	B734
737-4U3	B734
737-4Y0	B734
737-4Z6	B734
737-4Z9	B734
737-505	B735
737-522	B735
737-524	B735
737-528	B735
737-529	B735
737-530	B735
737-53A	B735
737-53C	B735
737-53S	B735
737-548	B735
737-54K	B735
737-55D	B735
737-55S	B735
737-566	B735
737-56N	B735
737-58E	B735
737-58N	B735
737-59D	B735
737-5B6	B735
737-5C9	B735
737-5H3	B735
737-5H4	B735
737-5H6	B735
737-5K5	B735
737-5L9	B735
737-5Q8	B735
737-5U3	B735
737-5Y0	B735
737-66N	B736
737-683	B736
737-6D6	B736
737-6H3	B736
737-6Q8	B736
737-6Z9	B736
737-705	B737
737-71M	B737
737-71Q	B737
737-724	B737
737-72T	B737
737-72U	B737
737-73A	B737
737-73Q	B737
737-73S	B737
737-73T	B737
737-73U	B737
737-73V	B737
737-74Q	B737
737-74T	B737
737-74T(BW)	B737
737-74U	B737
737-74V	B737
737-752	B737
737-752(BW)	B737
737-758	B737
737-75B	B737
737-75C	B737

737-75R	B737
737-75T	B737
737-75U	B737
737-75V	B737
737-75V(BW)	B737
737-760(BW)	B737
737-76D	B737
737-76N	B737
737-76N(BW)	B737
737-76Q	B737
737-76Q(BW)	B737
737-77L	B737
737-783	B737
737-78J	B737
737-78S	B737
737-790	B737
737-79K	B737
737-79L	B737
737-79P	B737
737-79T	B737
737-79U	B737
737-79U(BW)	B737
737-7AD	B737
737-7AF	B737
737-7AFC	B737
737-7AH	B737
737-7AH(BW)	B737
737-7AJ	B737
737-7AK	B737
737-7AN	B737
737-7AV	B737
737-7AW	B737
737-7AX	B737
737-7AXC	B737
737-7B6	B737
737-7BC	B737
737-7BC(BW)	B737
737-7BF	B737
737-7BH	B737
737-7BH(BW)	B737
737-7BJ	B737
737-7BK	B737
737-7BQ	B737
737-7BX	B737
737-7BX(BW)	B737
737-7CG	B737
737-7CJ	B737
737-7CN	B737
737-7CN(BW)	B737
737-7CP	B737
737-7CP(BW)	B737
737-7CT	B737
737-7CU	B737
737-7DF	B737
737-7DM	B737
737-7DP	B737
737-7DT	B737
737-7DW	B737
737-7E0	B737
737-7EA	B737
737-7ED	B737
737-7EG	B737
737-7EJ	B737
737-7EL	B737
737-7ET	B737
737-7FB(BW)	B737
737-7FG	B737
737-7H3	B737
737-7H4	B737
737-7H4(BW)	B737
737-7H6	B737
737-7H6(BW)	B737
737-7K2	B737
737-7K9	B737
737-7L9	B737
737-7P3	B737
737-7Q8	B737
737-7U8(BW)	B737
737-7V3	B737
737-7V3(BW)	B737
737-7W0	B737
737-7X2	B737
737-7Z5	B737
737-7Z5(BW)	B737
737-7Z9	B737
737-804	B738
737-809	B738
737-81B	B738
737-81M	B738
737-81Q	B738

737-81Q(BW)	B738
737-823	B738
737-824	B738
737-82R	B738
737-832	B738
737-838(BW)	B738
737-83N	B738
737-83N(BW)	B738
737-844(BW)	B738
737-84P	B738
737-84P(BW)	B738
737-858	B738
737-85F	B738
737-85F(BW)	B738
737-85H	B738
737-85P	B738
737-85R	B738
737-86D(BW)	B738
737-86J	B738
737-86J(BW)	B738
737-86N	B738
737-86N(BW)	B738
737-86Q	B738
737-86Q(BW)	B738
737-86R	B738
737-883	B738
737-89L	B738
737-8AN	B738
737-8AR	B738
737-8AS	B738
737-8B5	B738
737-8B6	B738
737-8BG	B738
737-8BG(BW)	B738
737-8BK	B738
737-8BK(BW)	B738
737-8CX	B738
737-8CX(BW)	B738
737-8D6	B738
737-8DP	B738
737-8DR	B738
737-8DV	B738
737-8EC	B738
737-8EQ	B738
737-8EV(BW)	B738
737-8EX(BW)	B738
737-8F2	B738
737-8F2(BW)	B738
737-8FE(BW)	B738
737-8K2	B738
737-8K5	B738
737-8K5(BW)	B738
737-8Q8	B738
737-8Q8(BW)	B738
737-8S3	B738
737-8S3(BW)	B738
737-8V3(BW)	B738
737-8X2	B738
737-8Z0	B738
737-8Z9	B738
737-8Z9(BW)	B738
737-924	B739
737-95R	B739
737-990	B739
737-9B5	B739
737-9K2	B739
737-T43A	B732
747-121	B741
747-121F	B741
747-122F	B741
747-123	B741
747-123F	B741
747-124	B741
747-124F	B741
747-128	B741
747-128F	B741
747-130	B741
747-131	B741
747-131F	B741
747-132F	B741
747-136	B741
747-143	B741
747-146	B741
747-146B	B741
747-146F	B741
747-148	B741
747-151	B741
747-168B	B741
747-186B	B741
747-206B	B742

747-206BF	B742
747-209B	B742
747-209BF	B742
747-209F	B742
747-211B	B742
747-212B	B742
747-212BF	B742
747-212F	B742
747-217B	B742
747-219B	B742
747-21AC	B742
747-221F	B742
747-222B	B742
747-227B	B742
747-228B	B742
747-228BF	B742
747-228F	B742
747-230B	B742
747-230BF	B742
747-230F	B742
747-233B	B742
747-236B	B742
747-236BF	B742
747-236F	B742
747-237B	B742
747-237BF	B742
747-238B	B742
747-238BF	B742
747-240B	B742
747-243B	B742
747-243BF	B742
747-243F	B742
747-244B	B742
747-244BF	B742
747-245F	B742
747-246B	B742
747-246BF	B742
747-246F	B742
747-249F	B742
747-251B	B742
747-251F	B742
747-256B	B742
747-256BF	B742
747-257B	B742
747-258B	B742
747-258C	B742
747-259BF	B742
747-267B	B742
747-267BF	B742
747-267F	B742
747-268F	B742
747-269B	B742
747-269BF	B742
747-270C	B742
747-271C	B742
747-273C	B742
747-281B	B742
747-281BF	B742
747-281F	B742
747-282B	B742
747-283B	B742
747-283BF	B742
747-284B	B742
747-286B	B742
747-287B	B742
747-2B2B	B742
747-2B3B	B742
747-2B3F	B742
747-2B4BF	B742
747-2B5B	B742
747-2B5BF	B742
747-2B5F	B742
747-2B6B	B742
747-2D3BF	B742
747-2D7BF	B742
747-2F6B	B742
747-2F6BF	B742
747-2G4B	B742
747-2H7B	B742
747-2J6B	B742
747-2J6BF	B742
747-2J6F	B742
747-2J9F	B742
747-2L5BF	B742
747-2Q2B	B742
747-2R7F	B742
747-2S4F	B742
747-2U3B	B742
747-2U3BF	B742
747-306	B743

747-312	B743
747-329	B743
747-329F	B743
747-337	B743
747-338	B743
747-341	B743
747-341F	B743
747-344	B743
747-346	B743
747-346SR	B743
747-357	B743
747-366	B743
747-367	B743
747-368	B743
747-3B3	B743
747-3B5	B743
747-3D7	B743
747-3G1	B743
747-3H6	B743
747-406	B744
747-406ERF	B74R
747-409	B744
747-409F	B744
747-412	B744
747-412F	B744
747-419	B744
747-41BF	B744
747-41R	B744
747-422	B744
747-428	B744
747-428ERF	B744
747-428F	B744
747-430	B744
747-433	B744
747-436	B744
747-437	B744
747-438	B744
747-438ER	B744
747-441	B744
747-443	B744
747-444	B744
747-446	B744
747-446D	B744
747-451	B744
747-458	B744
747-45E	B744
747-45EF	B744
747-467	B744
747-467F	B744
747-468	B744
747-469	B744
747-46NF	B744
747-475	B744
747-47C	B744
747-47UF	B744
747-481	B744
747-481D	B744
747-48E	B744
747-48EF	B744
747-4B3	B744
747-4B5	B744
747-4B5ERF	B744
747-4B5F	B744
747-4D7	B744
747-4F6	B744
747-4G4F	B744
747-4H6	B744
747-4J6	B744
747-4P8	B744
747-4Q8	B744
747-4R7F	B744
747-4U3	B744
747-E4B	B742
747SP-09	B74S
747SP-21	B74S
747SP-27	B74S
747SP-31	B74S
747SP-38	B74S
747SP-44	B74S
747SP-68	B74S
747SP-70	B74S
747SP-86	B74S
747SP-94	B74S
747SP-B5	B74S
747SP-J6	B74S
747SP-Z5	B74S
747SR-46	B74S
747SR-46F	B74R
747SR-81	B74R
747SR-81F	B74R

757-204	B752
757-208	B752
757-212	B752
757-21B	B752
757-21K	B752
757-222	B752
757-223	B752
757-224	B752
757-225	B752
757-22K	B752
757-22L	B752
757-230	B752
757-231	B752
757-232	B752
757-236	B752
757-236F	B752
757-23A	B752
757-23APF	B752
757-23N	B752
757-23P	B752
757-24APF	B752
757-24Q	B752
757-251	B752
757-256	B752
757-258	B752
757-25C	B752
757-25F	B752
757-260	B752
757-260PF	B752
757-26D	B752
757-27A	B752
757-27B	B752
757-28A	B752
757-28S	B752
757-29J	B752
757-2B6	B752
757-2B7	B752
757-2F8	B752
757-2F8C	B752
757-2G4	B752
757-2G5	B752
757-2G7	B752
757-2J4	B752
757-2K2	B752
757-2M6	B752
757-2Q8	B752
757-2S7	B752
757-2T7	B752
757-2Y0	B752
757-2Z0	B752
757-308	B753
757-324	B753
757-330	B753
757-33N	B753
757-351	B753
757-3CQ	B753
757-3E7	B753
767-201ER	B762
767-204	B762
767-204ER	B762
767-205	B762
767-209ER	B762
767-216ER	B762
767-219ER	B762
767-222	B762
767-223	B762
767-223ER	B762
767-224ER	B762
767-231	B762
767-232	B762
767-233	B762
767-233ER	B762
767-238ER	B762
767-23BER	B762
767-241ER	B762
767-246	B762
767-258	B762
767-258ER	B762
767-259ER	B762
767-25DER	B762
767-25E	B762
767-260ER	B762
767-266ER	B762
767-269ER	B762
767-275	B762
767-277	B762
767-27CER	B762
767-27EER	B762
767-27GER	B762
767-281	B762

767-283ER	B762
767-284ER	B762
767-29NER	B762
767-2AXER	B762
767-2B1ER	B762
767-2B7ER	B762
767-2DXER	B762
767-2J6ER	B762
767-2N0ER	B762
767-2Q4	B762
767-2Q8ER	B762
767-304ER	B763
767-306ER	B763
767-316ER	B763
767-316F	B763
767-319ER	B763
767-31AER	B763
767-31BER	B763
767-31KER	B763
767-322ER	B763
767-323ER	B763
767-324ER	B763
767-328ER	B763
767-330ER	B763
767-332	B763
767-332ER	B763
767-333ER	B763
767-336ER	B763
767-338ER	B763
767-33AER	B763
767-33PER	B763
767-341ER	B763
767-343ER	B763
767-346	B763
767-346ER	B763
767-34AF	B763
767-34PER	B763
767-352ER	B763
767-35DER	B763
767-35EER	B763
767-35HER	B763
767-360ER	B763
767-366ER	B763
767-36D	B763
767-36NER	B763
767-375ER	B763
767-37DER	B763
767-37EER	B763
767-381	B763
767-381ER	B763
767-381F	B763
767-383ER	B763
767-38AER	B763
767-38E	B763
767-38EER	B763
767-38EF	B763
767-39HER	B763
767-3BGER	B763
767-3CBER	B763
767-3D6	B763
767-3D6ER	B763
767-3G5ER	B763
767-3J6	B763
767-3P6ER	B763
767-3Q8ER	B763
767-3S1ER	B763
767-3T7ER	B763
767-3W0ER	B763
767-3X2ER	B763
767-3Y0ER	B763
767-3Z9ER	B763
767-424ER	B764
767-432ER	B764
777-206ER	B772
777-212ER	B772
777-21B	B772
777-21BER	B772
777-21H	B772
777-21HER	B772
777-222	B772
777-222ER	B772
777-223ER	B772
777-224ER	B772
777-228ER	B772
777-232ER	B772
777-236	B772
777-236ER	B772
777-243ER	B772
777-246	B772
777-246ER	B772

777-24QER	B772
777-258ER	B772
777-266ER	B772
777-267	B772
777-268ER	B772
777-269ER	B772
777-26KER	B772
777-281	B772
777-281ER	B772
777-289	B772
777-28EER	B772
777-2ANER	B772
777-2B5ER	B772
777-2D7	B772
777-2H6ER	B772
777-2J6	B772
777-2Q8ER	B772
777-2Z9ER	B772
777-312	B773
777-31H	B773
777-346	B773
777-367	B773
777-381	B773
777-3B5	B773
777-3D7	B773
A300B2-101	A30B
A300B2-103	A30B
A300B2-1C	A30B
A300B2-203	A30B
A300B2-203FF	A30B
A300B2K-3C	A30B
A300B4-103	A30B
A300B4-103F	A30B
A300B4-120	A30B
A300B4-203	A30B
A300B4-203F	A30B
A300B4-203FF	A30B
A300B4-220	A30B
A300B4-220FF	A30B
A300B4-2C	A30B
A300B4-601	A30B
A300B4-603	A30B
A300B4-605R	A30B
A300B4-620	A30B
A300B4-622	A30B
A300B4-622F	A30B
A300B4-622R	A30B
A300B4-622RF	A30B
A300C4-203	A306
A300C4-605R	A306
A300C4-620	A306
A300F4-203	A306
A300F4-605R	A306
A300F4-608ST	A300
A300F4-622R	A306
A310-203	A310
A310-203F	A310
A310-204	A310
A310-221F	A310
A310-222	A310
A310-222F	A310
A310-304	A310
A310-304F	A310
A310-308	A310
A310-322	A310
A310-324	A310
A310-324ET	A310
A310-324F	A310
A310-325	A310
A310-325ET	A310
A318-111	A318
A319-111	A319
A319-112	A319
A319-113	A319
A319-114	A319
A319-115X	A319
A319-131	A319
A319-132	A319
A319-132LR	A319
A319-132X	A319
A319-133LR	A319
A319-133X	A319
A320-111	A320
A320-211	A320
A320-212	A320
A320-214	A320
A320-231	A322
A320-232	A320
A320-233	A322
A321-111	A321

A321-111	A321
A321-112	A321
A321-131	A321
A321-211	A321
A321-211	A321
A321-231	A321
A321-232	A321
A330-201	A332
A330-202	A332
A330-203	A332
A330-223	A332
A330-243	A332
A330-301	A333
A330-303	A333
A330-321	A333
A330-322	A333
A330-323X	A333
A330-341	A333
A330-342	A333
A330-343X	A333
A340-211	A342
A340-212	A342
A340-213	A342
A340-311	A343
A340-312	A343
A340-313	A343
A340-313X	A343
A340-541	A345
A340-642	A346
AC-130H	C130
AC-130U	C130
AN-12	AN12
AN-124	A124
AN-12B	AN12
AN-12BK	AN12
AN-12BP	AN12
AN-12MGA	AN12
AN-12P	AN12
AN-12PPS	AN12
AN-12PS	AN12
AN-12RKR	AN12
AN-12TB	AN12
AN-12TBK	AN12
AN-12TBK-1	AN12
AN-12V	AN12
AN-140	A140
AN-22	AN22
AN-225	A225
AN-22A	AN22
AN-24	AN24
AN-24B	AN24
AN-24RT	AN24
AN-24RV	AN24
AN-24T	AN24
AN-24TV	AN24
AN-26	AN26
AN-26B	AN26
AN-26B-100	AN26
AN-26D	AN26
AN-26Z-1	AN26
AN-28	AN28
AN-30	AN30
AN-30D	AN30
AN-32	AN32
AN-32A	AN32
AN-32B	AN32
AN-38-100	AN38
AN-38-200	AN38
AN-72	AN72
AN-72-100D	AN72
AN-72P	AN72
AN-74	AN74
AN-74-200	AN74
AN-74T	AN74
AN-74T-100	AN74
AN-74T-200	AN74
AN-74TK	AN74
AN-74TK100	AN74
AN-74TK-100	AN74
AN-74TK-200	AN74
AN-74TK-300	AN74
AN-8	AN8
ANDOVERC 1	A748
ANDOVERC 1PR	A748
ANDOVERE 3	A748
ATL-98	CARV
ATP	ATP
ATR42-300	AT43
ATR42-300F	AT43
ATR42-320	AT43

ATR42-320F	AT43
ATR42-400MP	AT44
ATR42-420	AT44
ATR42-500	AT44
ATR42M-300	AT43
ATR72-102	AT72
ATR72-201	AT72
ATR72-202	AT72
ATR72-211	AT72
ATR72-212	AT72
ATR72-212A	AT72
BAC-111-201AC	BA11
BAC-111-203AE	BA11
BAC-111-204AF	BA11
BAC-111-208AL	BA11
BAC-111-211AH	BA11
BAC-111-212AR	BA11
BAC-111-401AK	BA11
BAC-111-402AP	BA11
BAC-111-407AW	BA11
BAC-111-408EF	BA11
BAC-111-409AY	BA11
BAC-111-410AQ	BA11
BAC-111-412EB	BA11
BAC-111-414EG	BA11
BAC-111-416EK	BA11
BAC-111-419EP	BA11
BAC-111-422EQ	BA11
BAC-111-423ET	BA11
BAC-111-476FM	BA11
BAC-111-479FU	BA11
BAC-111-481FW	BA11
BAC-111-485GD	BA11
BAC-111-487GK	BA11
BAC-111-488GH	BA11
BAC-111-492GM	BA11
BAC-111-501EX	BA11
BAC-111-509EW	BA11
BAC-111-510ED	BA11
BAC-111-515FB	BA11
BAC-111-518FG	BA11
BAC-111-520FN	BA11
BAC-111-521FH	BA11
BAC-111-523FJ	BA11
BAC-111-524FF	BA11
BAC-111-525FT	BA11
BAC-111-528FL	BA11
BAC-111-530FX	BA11
BAC-111-531FS	BA11
BAC-111-537GF	BA11
BAC-111-539GL	BA11
BAC-111-561RC	BA11
BAE 125-1000A	H25A
BAE 125-1000B	H25B
BAE146-100	BA46
BAE146-200	BA46
BAE146-200QC	BA46
BAE146-200QT	BA46
BAE146-300	BA46
BAE146-300	BA46
BAE146-300QT	BA46
BAE146-RJ100	BA46
BAE146-RJ70	BA46
BAE146-RJ85	BA46
BAE146-RJ85	BA46
BAE3100	JS31
BAE3101	JS31
BAE3102	JS31
BAE3103	JS31
BAE3107	JS31
BAE3108	JS31
BAE3109	JS31
BAE3112	JS31
BAE3116	JS31
BAE3200	JS32
BAE3201	JS32
BAE3201EP	JS32
BAE3202	JS32
BAE3202EP	JS32
BAE3206	JS32
BAE3212	JS32
BAE3212EP	JS32
BAE3217	JS32
BAE4100	JS41
BAE4101	JS41
BAE4102	JS41
BAE4107	JS41
BAE4112	JS41
BAE4121	JS41
BAE4124	JS41

BD-700-1A10	GLEX
BD-700-1A10(GLOBALEXP)	GLEX
BEECH1300	BE30
BEECH-1300	BE20
BEECH1900	B190
BEECH1900C	B190
BEECH1900C-1	B190
BEECH1900D	B190
BEECH-200	BE20
BEECH200	BE20
BEECH200C	BE20
BEECH-200C	BE20
BEECH-200T	BE20
BEECH200T	BE20
BEECH300	BE30
BEECH-300	BE30
BEECH300LW	BE30
BEECH-300LW	BE20
BEECH400	BE40
BEECH-400	BE40
BEECH400A	BE40
BEECH-400A	BE40
BEECH400T	BE40
BEECH-400T	BE40
BEECH99	BE99
BEECH99A	BE99
BEECHA200	BE20
BEECH-B200	BE20
BEECHB200	BE20
BEECH-B200C	BE20
BEECHB200C	BE20
BEECHB200CT	BE20
BEECH-B200CT	BE20
BEECHB200T	BE20
BEECH-B200T	BE20
BEECH-B300	BE30
BEECHB300	BE30
BEECHB300C	BE30
BEECH-B300C	BE30
BEECHB99	BE99
BEECHC-12J	BE20
BEECHC99	BE99
BN-2	BN2P
BN-2A	BN2P
BN-2A-2	BN2P
BN-2A-20	BN2P
BN-2A-21	BN2P
BN-2A-26	BN2P
BN-2A-26R	BN2P
BN-2A-27	BN2P
BN-2A-27LN	BN2P
BN-2A-27R	BN2P
BN-2A-3	BN2P
BN-2A-3S	BN2P
BN-2A-6	BN2P
BN-2A-6R	BN2P
BN-2A-7	BN2P
BN-2A-7R	BN2P
BN-2A-8	BN2P
BN-2A-8R	BN2P
BN-2A-9	BN2P
BN-2A-III	BN2P
BN-2A-III-1	BN2P
BN-2A-III-2	BN2P
BN-2A-III-3	BN2P
BN-2B-20	BN2P
BN-2B-21	BN2P
BN-2B-26	BN2P
BN-2B-27	BN2P
BN-2B-27R	BN2P
BN-2T	BN2T
BN-2T-4R	BN2T
BN-2T-4S	BN2T
C-118A	DC6
C-118B	DC6
C-121C	CONI
C-121G	CONI
C-121J	CONI
C-12C	BE20
C-12D	BE20
C-12F	BE20
C-12L	BE20
C-12R	BE20
C-130A	C130
C-130B	C130
C-130E	C130
C-130H	C130
C-130H-30	C130
C-130J	C130
C-130J-30	C130

C-130K-C 1	C130
C-130K-C 3	C130
C-130K-W 2	C130
C-130T	C130
C-131A	CVLT
C-131B	CVLT
C-131D	CVLT
C-131E	CVLT
C-131F	CVLT
C-131H	CVLT
C-20A	GLF3
C-20B	GLF3
C-20C	GLF3
C-20D	GLF3
C-20E	GLF3
C-20F	GLF3
C-20G	GLF3
C-20H	GLF3
C-21A	LJ35
C-23A	SH36
C-23B	SH33
C-23B+	SH33
C-29A	H25A
C-37A	GLF5
C-38A	WW25
C-46A-45-CU	C46
C-46A-50-CU	C46
C-46A-55-CK	C46
C-46A-5-CU	C46
C-46A-60-CK	C46
C-46A-60-CS	C46
C-46D-10-CU	C46
C-46F	C46
C-46F-1-CU	C46
C-46R	C46
C-54A-10-DC	DC4
C-54A-15-DC	DC4
C-54A-1-DC	DC4
C-54A-1-DO	DC4
C-54A-5-DO	DC4
C-54A-DO	DC4
C-54B-10-DO	DC4
C-54B-15-DO	DC4
C-54B-1-DC	DC4
C-54B-5-DO	DC4
C-54D-10-DC	DC4
C-54D-15-DC	DC4
C-54D-1-DC	DC4
C-54D-5-DC	DC4
C-54E-15-DO	DC4
C-54E-1-DO	DC4
C-54E-5-DO	DC4
C-54G-10-DO	DC4
C-54G-15-DO	DC4
C-54G-1-DO	DC4
C-54G-5-DO	DC4
C-54M	DC4
C-54M-10-DO	DC4
C-54M-15-DO	DC4
C-54P	DC4
C-54Q	DC4
C-54R	DC4
C-54S	DC4
C-54T	DC4
C-97G	C97
C-9A	DC9
C-9B	DC9
CARAVELLE10B3	S210
CARAVELLE10R	S210
CARAVELLE11R	S210
CARAVELLE12	S210
CARAVELLE3	S210
CARAVELLE6R	S210
CASA212	C212
CASA212-100	C212
CASA212-200	C212
CASA212-300	C212
CASA212-400	C212
CASA212M-100	C212
CASA212M-200	C212
CASA212M-300	C212
CASA212MP-200	C212
CASA295	CN35
CESSNA208	C208
CESSNA208A	C208
CESSNA208B	C208
CESSNA500	C500
CESSNA-500	C500
CESSNA501	C501
CESSNA-501	C501

CESSNA525	C525
CESSNA-525	C525
CESSNA525A	C525
CESSNA-525A	C525
CESSNA-550	C550
CESSNA550	C550
CESSNA551	C551
CESSNA-551	C551
CESSNA-552	C551
CESSNA560	C560
CESSNA-560	C560
CESSNA560XL	C560
CESSNA-560XL	C560
CESSNA650	C650
CESSNA-650	C650
CESSNA750	C750
CESSNA-750	C750
CESSNAS550	C550
CESSNA-S550	C550
CL-44-6	CL44
CL-44D4-2	CL44
CL-44-O	CL44
CL-600-1A11	CL60
CL-600-1A11(CHALL 600)	CL60
CL-600-1A11-S	CL60
CL-600-1A11-SCHALL 600S	CL60
CL-600-2A12	CL60
CL-600-2A12(CHALL 601)	CL60
CL-600-2B16	CL60
CL-600-2B16CHALL 601-3A	CL60
CL-600-2B16CHALL 601-3R	CL60
CL-600-2B19	CL60
CL-600-2B19(CHALL 800)	CL60
CL-600-2B19(CRJ100)	CL60
CL-600-2B19(CRJ200)	CL60
CL-600-2B19(CRJ200ER)	CL60
CL-600-2B19(CRJ200LR)	CL60
CL-600-2B19(CRJ440)	CL60
CL-600-2C10	CL60
CL-600-2D24(CRJ900)	CL60
CL-604	CL60
CL-604(CHALLENGER604)	CL60
CL-66B	CL60
CN-235-1	CN35
CN-235-10	CN35
CN-235-100	CN35
CN-235-100MPA	CN35
CN-235-100QC	CN35
CN-235-200	CN35
CN-235-200QC	CN35
CN-235-220	CN35
CN-235-300	CN35
CN-235M-10	CN35
CN-235M-100	CN35
CN-235M-200	CN35
CN-235M-220	CN35
CN-235MP-100M	CN35
CN-235MPA	CN35
COMET4C	COMT
CONCORDE100	CONC
CONCORDE101	CONC
CONCORDE102	CONC
CT-29A	CVLP
CT-39A	SBR1
CT-39E	SBR1
CT-39G	SBR1
CV-240-0	CVLP
CV-240-11	CVLP
CV-240-14	CVLP
CV-240-15	CVLP
CV-240-3	CVLP
CV-240-4	CVLP
CV-240-5	CVLP
CV-300	CVLP
CV-340-32	CVLP
CV-340-68B	CVLP
CV-440	CVLP
CV-440-0	CVLP
CV-440-11	CVLP
CV-440-38	CVLP
CV-440-40	CVLP
CV-440-58	CVLP
CV-440-62	CVLP
CV-440-75	CVLP
CV-440-78	CVLP
CV-440-80	CVLP
CV-440-86	CVLP
CV-440-94	CVLP
CV-440-98	CVLP
CV-540	CVLT

CV-580	CVLT
CV-5800	CVLT
CV-600	CVLT
CV-640	CVLT
CV-880-22-1	CV99
DC-10-10	DC10
DC-10-10CF	DC10
DC-10-10F	DC10
DC-10-15	DC10
DC-10-30	DC10
DC-10-30CF	DC10
DC-10-30ER	DC10
DC-10-30F	DC10
DC-10-40	DC10
DC-10-40F	DC10
DC-130A	C130
DC-4	DC4
DC-4-1009	DC4
DC-6	DC6
DC-6A	DC6
DC-6A/C	DC6
DC-6A\B	DC6
DC-6A\C	DC6
DC-6B	DC6
DC-6BF	DC6
DC-7	DC7
DC-7B	DC7
DC-7BF	DC7
DC-7C	DC7
DC-7CF	DC7
DC-8-51	DC85
DC-8-52	DC85
DC-8-53	DC85
DC-8-54F	DC85
DC-8-55F	DC85
DC-8-61	DC86
DC-8-61F	DC86
DC-8-62	DC86
DC-8-62F	DC86
DC-8-63	DC86
DC-8-63F	DC86
DC-8-71F	DC87
DC-8-72	DC87
DC-8-72F	DC87
DC-8-73F	DC87
DC-9-14	DC91
DC-9-15	DC91
DC-9-15F	DC9
DC-9-21	DC92
DC-9-31	DC93
DC-9-32	DC93
DC-9-32F	DC93
DC-9-33F	DC93
DC-9-34	DC93
DC-9-34F	DC93
DC-9-41	DC94
DC-9-51	DC95
DC-9-81	DC98
DC-9-82	DC98
DC-9-83	DC98
DC-9-87	DC98
DHC-6-100	DHC6
DHC-6-200	DHC6
DHC-6-300	DHC6
DHC-6-310	DHC6
DHC-6-320	DHC6
DHC-7-100	DHC7
DHC-7-102	DHC7
DHC-7-103	DHC7
DHC-7-110	DHC7
DHC-7-150	DHC7
DHC-8-101	DH8A
DHC-8-102	DH8A
DHC-8-102A	DH8A
DHC-8-103	DH8A
DHC-8-103A	DH8A
DHC-8-103B	DH8A
DHC-8-106	DH8A
DHC-8-110	DH8A
DHC-8-201	DH8B
DHC-8-202	DH8B
DHC-8-301	DH8C
DHC-8-311	DH8C
DHC-8-311A	DH8C
DHC-8-311B	DH8C
DHC-8-314	DH8C
DHC-8-314A	DH8C
DHC-8-314B	DH8C
DHC-8-315	DH8C
DHC-8-315B	DH8C

DHC-8Q-103	DH8A
DHC-8Q-103B	DH8A
DHC-8Q-201	DH8B
DHC-8Q-201B	DH8B
DHC-8Q-202	DH8B
DHC-8Q-202B	DH8B
DHC-8Q-311	DH8C
DHC-8Q-311B	DH8C
DHC-8Q-314	DH8C
DHC-8Q-314B	DH8C
DHC-8Q-315	DH8C
DHC-8Q-401	DH8D
DHC-8Q-402	DH8D
DO228-	D228
DO228-100	D228
DO228-101	D228
DO228-200	D228
DO228-201	D228
DO228-201K	D228
DO228-202	D228
DO228-202K	D228
DO228-212	D228
DO328-110	D328
DO328-120	D328
DO328-130	D328
DO328-300	D328
DO328-310	D328
DOMINIE-T 1	H25A
E-9A	DH8A
EC-121T	CONI
EC-130E	C130
EC-130H	C130
EC-130J	C130
EC-130Q	C130
EC-18B	B703
EC-18D	B703
EMB-110	E110
EMB-110A	E110
EMB-110B	E110
EMB-110B1	E110
EMB-110C	E110
EMB-110C\N	E110
EMB-110E	E110
EMB-110E\J	E110
EMB-110K1	E110
EMB-110P	E110
EMB-110P1	E110
EMB-110P1A	E110
EMB-110P1K	E110
EMB-110P2	E110
EMB-110S1	E110
EMB-111A	E110
EMB-111A\N	E110
EMB-120	E120
EMB-120ER	E120
EMB-120FC	E120
EMB-120QC	E120
EMB-120RT	E120
EMB-135BJ	E135
EMB-135ER	E135
EMB-135KL	E135
EMB-135LR	E135
EMB-135LR(SL)	E135
EMB-145	E145
EMB-145EP	E145
EMB-145ER	E145
EMB-145EU	E145
EMB-145LR	E145
EMB-145LU	E145
EMB-145MP	E145
EMB-145MR	E145
EMB-145RS	E145
EMB-145SA	E145
EMB-145XR	E145
F-27	F27
F-27-050	F27
F-27-0502	F27
F-27-050-300	F27
F-27-050S	F27
F-27-0604	F27
F-27-100	F27
F-27-200	F27
F-27-200MAR	F27
F-27-300	F27
F-27-300M	F27
F-27-400	F27
F-27-400M	F27
F-27-500	F27
F-27-500C	F27
F-27-500F	F27

F-27-500RF	F27
F-27-500RFC	F27
F-27-600	F27
F-27-600P	F27
F-27-700	F27
F-27A	F27A
F-27B	F27
F-27F	F27
F-27J	F27
F-28-0070	F28
F-28-0100	F28
F-28-1000	F28
F-28-1000C	F28
F-28-2000	F28
F-28-3000	F28
F-28-3000C	F28
F-28-3000M	F28
F-28-3000R	F28
F-28-3000RC	F28
F-28-4000	F28
FALCON10	FA10
FALCON-10	FA10
FALCON100	FA10
FALCON-100	FA10
FALCON-10MER	FA10
FALCON10VF	FA10
FALCON-10VF	FA10
FALCON200	FA20
FALCON-200	FA20
FALCON2000	F2TH
FALCON-2000	F2TH
FALCON2000EX	F2TH
FALCON20C	FA20
FALCON-20C	FA20
FALCON20C-5	FA20
FALCON-20C-5	FA20
FALCON-20C-5B	FA20
FALCON20D	FA20
FALCON-20D	FA20
FALCON-20D-5	FA20
FALCON20DC	FA20
FALCON-20DC	FA20
FALCON-20E	FA20
FALCON20E-5	FA20
FALCON-20E-5B	FA20
FALCON20EC	FA20
FALCON-20EC	FA20
FALCON-20ECM	FA20
FALCON20F	FA20
FALCON-20F	FA20
FALCON20F-5	FA20
FALCON-20F-5	FA20
FALCON20F-5B	FA20
FALCON-20F-5B	FA20
FALCON20F-5BR	FA20
FALCON-20G	FA20
FALCON20GF-5	FA20
FALCON-20SNA	FA20
FALCON50	FA50
FALCON-50	FA50
FALCON-50-40	FA50
FALCON50EX	FA50
FALCON-50EX	FA50
FALCON900	F900
FALCON-900	F900
FALCON900B	F900
FALCON-900B	F900
FALCON900C	F900
FALCON-900C	F900
FALCON900EX	F900
FALCON-900EX	F900
FALCON900EXEASY	F900
FH-227	F27
FH-227B	F27
FH-227D	F27
FH-227E	F27
G-1159	GLF2
G-1159A	GLF3
G-1159A-SRA-1	GLF3
G-1159B	GLF3
G-1159SP	GLF4
G-1159-SP	GLF2
G-1159TT	GLF2
G-159	G159
G-159C	G159
G-IV	GLF4
G-IVSP	GLF4
GULFSTREAM100	G159
GULFSTREAM200	GLF2
GULFSTREAM-200	GLF2

GULFSTREAM300	GLF3
GULFSTREAM400	GLF4
GULFSTREAM550	GLF5
G-V	GLF5
HC-130H	C130
HC-130H-7	C130
HC-130J	C130
HC-130N	C130
HC-130P	C130
HC-131A	CVLT
HC-54D-10DC	DC4
HERON-1B	HERN
HERON-2	HERN
HERON-2B	HERN
HERON-2C	HERN
HERON-2D	HERN
HFB 320	HF20
HP 137-200	JS20
HP 137-III	JS20
HP 137MK 1	JS20
HP 137TMK 1	JS20
HP 137TMK 2	JS20
HS 125-1A-521	H25A
HS 125-1A-522	H25A
HS 125-1AS-522	H25A
HS 125-1B-522	H25A
HS 125-1BS-522	H25A
HS 125-3A	H25A
HS 125-3A/RA	H25A
HS 125-3A\R	H25A
HS 125-3A\RA	H25A
HS 125-3B	H25A
HS 125-3B\RA	H25A
HS 125-3B\RC	H25A
HS 125-400A	H25A
HS 125-400B	H25A
HS 125-403B	H25A
HS 125-600A	H25A
HS 125-600B	H25A
HS 125-700A	H25B
HS 125-700A/TR	H25B
HS 125-700A\TR	H25B
HS 125-700B	H25B
HS 125-800A	H25A
HS 125-800B	H25A
HS 125-800RA	H25A
HS 125-800SIG	H25A
HS 125-800XP	H25B
HS 125-800XRA	H25A
HS 125-F1A	H25A
HS 125-F3A	H25A
HS 125-F3A\RA	H25A
HS 125-F3B	H25A
HS 125-F3B\RA	H25A
HS 125-F400A	H25A
HS 125-F400B	H25A
HS 125-F403B	H25A
HS 125-F600A	H25A
HS 125-F600B	H25A
HS 748-1\103	A748
HS 748-1\104	A748
HS 748-1\105	A748
HS 748-1\107	A748
HS 748-2\203	A748
HS 748-2\204	A748
HS 748-2\205	A748
HS 748-2\206	A748
HS 748-2\207	A748
HS 748-2\208	A748
HS 748-2\209	A748
HS 748-2\218	A748
HS 748-2\219	A748
HS 748-2\220	A748
HS 748-2\224	A748
HS 748-2\228	A748
HS 748-2\229	A748
HS 748-2\232	A748
HS 748-2\243	A748
HS 748-2\244	A748
HS 748-2\247	A748
HS 748-2\268	A748
HS 748-2A/229	A748
HS 748-2A/242	A748
HS 748-2A/244	A748
HS 748-2A/264	A748
HS 748-2A/269	A748
HS 748-2A/286	A748
HS 748-2A/333	A748
HS 748-2A/351	A748
HS 748-2A\209	A748

Aircraft	Code
HS 748-2A\210	A748
HS 748-2A\214	A748
HS 748-2A\215	A748
HS 748-2A\216	A748
HS 748-2A\221	A748
HS 748-2A\225	A748
HS 748-2A\229	A748
HS 748-2A\233	A748
HS 748-2A\234	A748
HS 748-2A\235	A748
HS 748-2A\238	A748
HS 748-2A\242	A748
HS 748-2A\244	A748
HS 748-2A\245	A748
HS 748-2A\246	A748
HS 748-2A\248	A748
HS 748-2A\253	A748
HS 748-2A\254	A748
HS 748-2A\256	A748
HS 748-2A\257	A748
HS 748-2A\258	A748
HS 748-2A\263	A748
HS 748-2A\264	A748
HS 748-2A\265	A748
HS 748-2A\266	A748
HS 748-2A\267	A748
HS 748-2A\269	A748
HS 748-2A\270	A748
HS 748-2A\271	A748
HS 748-2A\272	A748
HS 748-2A\273	A748
HS 748-2A\274	A748
HS 748-2A\275	A748
HS 748-2A\276	A748
HS 748-2A\281	A748
HS 748-2A\282	A748
HS 748-2A\283	A748
HS 748-2A\285	A748
HS 748-2A\286	A748
HS 748-2A\288	A748
HS 748-2A\301	A748
HS 748-2A\309	A748
HS 748-2A\310	A748
HS 748-2A\320	A748
HS 748-2A\333	A748
HS 748-2A\334	A748
HS 748-2A\335	A748
HS 748-2A\343	A748
HS 748-2A\344	A748
HS 748-2A\347	A748
HS 748-2A\351	A748
HS 748-2A\369	A748
HS 748-2A\372	A748
HS 748-2B/371	A748
HS 748-2B/399	A748
HS 748-2B/426	A748
HS 748-2B/FAA	A748
HS 748-2B\287	A748
HS 748-2B\360	A748
HS 748-2B\371	A748
HS 748-2B\378	A748
HS 748-2B\398	A748
HS 748-2B\399	A748
HS 748-2B\400	A748
HS 748-2B\401	A748
HS 748-2B\402	A748
HS 748-2B\424	A748
HS 748-2B\426	A748
HS 748-2B\435	A748
HS 748-2B\501	A748
HS 748-2B\FAA	A748
HS 748-2M	A748
HU-25A	FA20
HU-25B	FA20
HU-25C	FA20
HU-25D	FA20
IAI-101-ARAVA	ARVA
IAI-101B-ARAVA	ARVA
IAI102ARAVA	ARVA
IAI-102-ARAVA	ARVA
IAI1123	WW24
IAI-1123	WW24
IAI-1123N	WW24
IAI-1123QC	WW24
IAI1124	WW24
IAI-1124	WW24
IAI1124A	WW24
IAI-1124A	WW24
IAI-1124N	WW24
IAI1125	ASTR

IAI-1125	ASTR
IAI1125SP	ASTR
IAI-1125SP	ASTR
IAI1125SPX	ASTR
IAI-1125SPX	ASTR
IAI1126	WW24
IAI-1126	WW24
IAI-201-ARAVA	ARVA
IAI-202-ARAVA	ARVA
IL-114	I114
IL-114-100	I114
IL-18	IL18
IL-18B	IL18
IL-18D	IL18
IL-18E	IL18
IL-18GRM	IL18
IL-18V	IL18
IL-22M	IL18
IL-62	IL62
IL-62M	IL62
IL-62MK	IL62
IL-76	IL76
IL-76\976	IL76
IL-76\A50	IL76
IL-76M	IL76
IL-76MD	IL76
IL-76MDK	IL76
IL-76T	IL76
IL-76TD	IL76
IL-76VPK	IL76
IL-78	IL76
IL-78M	IL76
IL-78MD	IL76
IL-86	IL86
IL-96-300	IL96
JC-12C	BE20
JC-12D	BE20
JETSTAR-6	L29A
JETSTAR731	L29B
JETSTAR-731	L29B
JETSTAR-8	L29A
JETSTARII	L29B
JETSTAR-II	L29B
KC-10A	DC10
KC-130B	C130
KC-130F	C130
KC-130H	C130
KC-130J	C130
KC-130R	C130
KC-130T	C130
KC-130T-30	C130
L-100	C130
L-100-20	C130
L-100-30	C130
L-1011-1	L101
L-1011-100	L101
L-1011-150	L101
L-1011-1F	L101
L-1011-200	L101
L-1011-200F	L101
L-1011-250	L101
L-1011-385-1(1)	L101
L-1011-385-1(50)	L101
L-1011-385-1-15(100)	L101
L-1011-385-1-15(250)	L101
L-1011-385-1-15F(200F)	L101
L-1011-385-3(500)	L101
L-1011-50	L101
L-1011-500	L101
L-1011-500F	L101
L-1049H	CONI
L-1649A-98	CONI
L-188A	L188
L-188AF	L188
L-188C	L188
L-188CF	L188
L-188PF	L188
L410A	L410
L410AB	L410
L410AS	L410
L410FG	L410
L410IVP-4	L410
L410M	L410
L410MA	L410
L410MU	L410
L410T	L410
L410UVO-E10	L410
L410UVP	L410
L410UVP-4	L410
L410UVP-E	L410

L410UVP-E1	L410
L410UVP-E10	L410
L410UVP-E12	L410
L410UVP-E13	L410
L410UVP-E14	L410
L410UVP-E15	L410
L410UVP-E16	L410
L410UVP-E17	L410
L410UVP-E18	L410
L410UVP-E19	L410
L410UVP-E2	L410
L410UVP-E20	L410
L410UVP-E20C	L410
L410UVP-E20D	L410
L410UVP-E20G	L410
L410UVP-E20K	L410
L410UVP-E3	L410
L410UVP-E4	L410
L410UVP-E5	L410
L410UVP-E6	L410
L410UVP-E7	L410
L410UVP-E8	L410
L410UVP-E8A	L410
L410UVP-E8B	L410
L410UVP-E8C	L410
L410UVP-E8D	L410
L410UVP-E8E	L410
L410UVP-E9	L410
L410UVP-E9D	L410
L410X	L410
L420	L410
L610M	L610
L-749A-79-36	CONI
L-749A-79-38	CONI
L-749A-79-43	CONI
LC-130F	C130
LC-130H	C130
LC-130R	C130
LEARJET23	LJ23
LEARJET-23	LJ23
LEARJET24	LJ24
LEARJET-24	LJ24
LEARJET-24A	LJ24
LEARJET24B	LJ24
LEARJET-24B	LJ24
LEARJET24D	LJ24
LEARJET-24D	LJ24
LEARJET24E	LJ24
LEARJET-24E	LJ24
LEARJET-24F	LJ24
LEARJET24XR	LJ24
LEARJET-24XR	LJ24
LEARJET25	LJ25
LEARJET-25	LJ25
LEARJET25B	LJ25
LEARJET-25B	LJ25
LEARJET-25C	LJ25
LEARJET25D	LJ25
LEARJET-25D	LJ25
LEARJET25XR	LJ25
LEARJET-25XR	LJ25
LEARJET28	LJ28
LEARJET-28	LJ28
LEARJET-29	LJ29
LEARJET31	LJ31
LEARJET-31	LJ31
LEARJET31A	LJ31
LEARJET-31A	LJ31
LEARJET35	LJ35
LEARJET-35	LJ35
LEARJET35A	LJ35
LEARJET-35A	LJ35
LEARJET36	LJ36
LEARJET-36	LJ36
LEARJET36A	LJ36
LEARJET-36A	LJ36
LEARJET45	LJ45
LEARJET-45	LJ45
LEARJET45XR	LJ45
LEARJET55	LJ55
LEARJET-55	LJ55
LEARJET55B	LJ55
LEARJET-55B	LJ55
LEARJET55C	LJ55
LEARJET-55C	LJ55
LEARJET60	LJ60
LEARJET-60	LJ60
M-28	M28
M-28B1	M28
M-28B1R	M28

M-28B1TD	M28
M-28BTD2	M28
MARTIN4-0-4	M404
MARTIN-4-0-4	M404
MC-130E	C130
MC-130E-Y	C130
MC-130H	C130
MC-130P	C130
MD-10-10	DC10
MD-10-10F	DC10
MD-10-30CF	DC10
MD-10-30F	DC10
MD-11	MD11
MD-11C	MD11
MD-11CF	MD11
MD-11ER	MD11
MD-11F	MD11
MD-88	MD83
MD-88	MD83
MD-90-30	MD93
MINI-GUPPY	SGUP
MU-300-1	MU30
MU-300-1A	MU30
MU-300-2	MU30
N22B	NOMA
N22BFP	NOMA
N22C	N22
N22P	N22
N22SB	N22
N22SL	N22
N24A	NOMA
NA-265-40(SABRE40)	SBR1
NA-265-40(SABRE40A)	SBR1
NA-265-60(SABRE60)	SBR1
NA-265-60(SABRE60SC)	SBR1
NA-265-60(SABRE65)	SBR1
NA-265-65(SABRE65)	SBR1
NA-265-80(SABRE75A)	SBR1
NC-130A	C130
NC-130B	C130
NC-130E	C130
NC-130H	C130
NC-131H	CVLT
NORD262	N262
NORD262A	N262
NORD262A12	N262
NORD262A14	N262
NORD262A20	N262
NORD262A21	N262
NORD262A26	N262
NORD262A27	N262
NORD262A28	N262
NORD262A29	N262
NORD262A32	N262
NORD262A34	N262
NORD262A36	N262
NORD262A37	N262
NORD262A38	N262
NORD262A40	N262
NORD262A41	N262
NORD262A42	N262
NORD262A44	N262
NORD262A45	N262
NORD262AG43	N262
NORD262C5OP	N262
NORD262C61	N262
NORD262C62	N262
NORD262C63	N262
NORD262C64	N262
NORD262C65	N262
NORD262C67	N262
NORD262CS	N262
NORD262D51	N262
NT-39A	SBR1
PD-808GE1	P808
PD-808GE2	P808
PD-808TA	P808
PD-808TP	P808
RC-12D	BE20
RC-12G	BE20
RC-12H	BE20
RC-12K	BE20
RC-12N	BE20
RC-12P	BE20
RC-12Q	BE20
RILEYHERON	HERN
RILEY-HERON	HERN
S VC-10-K 3	VC10
S VC-10-K 4	VC10
SA226AT	SA26

SA226T	SA26
SA226T(A)	SA26
SA226T(B)	SA26
SA226TC	SA26
SA227AC	SA27
SA227AT	SA27
SA227BC	SA27
SA227CC	SA27
SA227DC	SA27
SA227TT	SA27
SAAB2000	SF20
SAAB340A	SF34
SAAB340B	SF34
SABRE-40	SBR1
SABRE-40A	SBR1
SABRE-40EX	SBR1
SABRE-40R	SBR1
SABRE-50	SBR1
SABRE-60	SBR1
SABRE-60A	SBR1
SABRE-60SC	SBR1
SABRE-65	SBR1
SABRE-75	SBR1
SABRE-75A	SBR1
SABRE-80A	SBR1
SC5	BELF
SC-54G-10DO	DC4
SC7-3	SC7
SC7-3-100	SC7
SC7-3-200	SC7
SC7-3A-100	SC7
SC7-3A-200	SC7
SC7-3M-100	SC7
SC7-3M-400	SC7
SD3-30	SH33
SD3-30-100	SH33
SD3-30-200	SH33
SD3-30-UTT	SH33
SD3-60-100	SH36
SD3-60-200	SH36
SD3-60-300	SH36
SN 601	S601
SN 601(CORVETTE)	S601
SUPER-46C	C46
SUPER-GUPPY201	SGUP
T-29B	CVLP
T-29C	CVLP
T-29D	CVLP
T-39A	SBR1
T-39B	SBR1
T-39D	SBR1
TC-12B	BE20
TC-130G	C130
TC-130Q	C130
TC-18E	B703
TC-18F	B703
TC-4C	G159
TU-134	T134
TU-134A	T134
TU-134A-1	T134
TU-134A-3	T134
TU-134A-3CX	T134
TU-134A-3M	T134
TU-134B-3	T134
TU-134SH	T134
TU-134UB	T134
TU-134UBL	T134
TU-154A	T154
TU-154B	T154
TU-154B-1	T154
TU-154B-2	T154
TU-154M	T154
TU-154M-LK1	T154
TU-154S	T154
TU-204	T204
TU-204-100S	T204
TU-204-120	T204
TU-204-120C	T204
TU-204-200C3	T204
TU-204C	T204
U-125	H25B
U-125A	H25B
U-36A	LJ36
U-4	GLF4
UC-12B	BE20
UC-12F	BE20
UC-12M	BE20
UC-35A	C560
UC-35C	C560
UC-35D	C560

VC-10-C 1K	VC10
VC-10-K 2	VC10
VC-118B	DC6
VC-130H	C130
VC-131D	CVLT
VC-4A	G159
VC-54N	DC4
VC-54Q	DC4
VC-9C	DC93
VFW-614	VF14
VISCOUNT798D	VISC
VISCOUNT802	VISC
VISCOUNT806	VISC
VISCOUNT836	VISC
VT-29B	CVLP
VT-29D	CVLP
WC-130E	C130
WC-130H	C130
WC-130J	C130
Y12-2	Y12
Y12-2S	Y12
Y12-4	Y12
Y7	YN7
Y7-100	YN7
Y7-100C	YN7
Y7-200	YN7
Y7H-200	AN26
Y8	AN12
Y8D	AN12
Y8F-100	AN12
Y8F-200	AN12
YAK-142	YK42
YAK-40	YK40
YAK-40D	YK40
YAK-40EC	YK40
YAK-40FG	YK40
YAK-40K	YK40
YAK-42	YK42
YAK-42D	YK42
YAK-42F	YK42
YS-11-101	YS11
YS-11-102	YS11
YS-11-103	YS11
YS-11-105	YS11
YS-11-108	YS11
YS-11-109	YS11
YS-11-110	YS11
YS-11-112	YS11
YS-11-113	YS11
YS-11-115	YS11
YS-11-117	YS11
YS-11-118	YS11
YS-11-120	YS11
YS-11-124	YS11
YS-11A-202	YS11
YS-11A-205	YS11
YS-11A-206	YS11
YS-11A-207	YS11
YS-11A-208	YS11
YS-11A-209	YS11
YS-11A-212	YS11
YS-11A-213	YS11
YS-11A-214	YS11
YS-11A-217	YS11
YS-11A-218	YS11
YS-11A-220	YS11
YS-11A-222	YS11
YS-11A-223	YS11
YS-11A-227	YS11
YS-11A-305	YS11
YS-11A-306	YS11
YS-11A-309	YS11
YS-11A-313	YS11
YS-11A-314	YS11
YS-11A-325	YS11
YS-11A-402	YS11
YS-11A-404	YS11
YS-11A-500	YS11
YS-11A-600	YS11
YS-11A-607	YS11
YS-11A-621	YS11
YS-11A-623	YS11
YS-11A-624	YS11
YS-11A-626	YS11
YS-11A-659	YS11

APPENDIX F: Engine Mappings

The following table provides a mapping of all unique engines from BACK to those from ICAO or EDMS in order to model emissions. In general, the ICAO mappings cover jet engines while the EDMS mappings cover turboprops and pistons. The mappings range from exact matches to approximate substitutions.

BACK Engine Name	ICAO UID	ICAO Engine Name	EDMS Engine Name
749C18-BD1			R-1820
988TC18-EA2			R-1820
988TC18-EA3			R-1820
AE2100A	4AL003	AE3007A	
AE-2100D3	4AL003	AE3007A	
AE2100D3	4AL003	AE3007A	
AE3007A	4AL003	AE3007A	
AE3007A1	6AL006	AE3007A1	
AE3007A1/1	6AL009	AE3007A1/1	
AE3007A1/2	4AL002	AE3007A1 series	
AE3007A1/3	6AL012	AE3007A1/3	
AE3007A1\1	4AL002	AE3007A1 series	
AE3007A1\2	4AL002	AE3007A1 series	
AE3007A1E	6AL020	AE3007A1E	
AE3007A1P	6AL015	AE3007A1P	
AE3007A3	6AL018	AE3007A3	
AE-3007C	6AL018	AE3007A3	
AE3007C	6AL018	AE3007A3	
AE-3007C-1	6AL018	AE3007A3	
AE3007C-1	6AL018	AE3007A3	
AI-20			501D22A
AI-20D			501D22A
AI-20M			501D22A
AI-24A			PW127-A
AI-24T			PW127-A
AI-24UT			PW127-A
AI-25			PW127-A
ALF502L-2C	1TL002	ALF 502R-3	
ALF502R-3	1TL002	ALF 502R-3	
ALF502R-3A	1TL002	ALF 502R-3	
ALF502R-5	1TL003	ALF 502R-5	
ALF502R-5-10	1TL003	ALF 502R-5	
ASN250-B17B			250B17B
ASN-250-B17C			250B17B
ASN250-B17C			250B17B
ASN-250-B17FI			250B17B
ASN501-D13A			501D22A
ASN501-D13D			501D22A
ASN501-D13H			501D22A
ASN501-D15			501D22A
ASN-501-D22			501D22A
ASN-501-D22A			501D22A
ASN501-D22A			501D22A
ASN-501-D22C			501D22A
ASN501-D22G			501D22A
ASN501-D36			501D22A
ASTAZOUXIV C1			TPE331-3
ASTAZOUXVI			TPE331-3
ASTAZOUXVI C1			TPE331-3
ASTAZOUXVI D			TPE331-3
ATF3-6A-4C			CF700-2D
AVON525B	1PW007	JT8D-9 series	
AVON527B	1PW007	JT8D-9 series	
AVON533R	1PW007	JT8D-9 series	
BASTANVIC			PT6A-27
BASTANVII			PT6A-27
BR710A1-10	4BR005	BR700-715A1-30	
BR710A2-20	4BR005	BR700-715A1-30	
BR710C4-11	4BR005	BR700-715A1-30	
BR715A1-30	4BR005	BR700-715A1-30	
BR715C1-30	4BR007	BR700-715C1-30	
CF34-1A	1GE035	CF34-3A1	
CF34-3A	1GE035	CF34-3A1	
CF34-3A1	1GE035	CF34-3A1	
CF34-3A2	1GE035	CF34-3A1	
CF34-3B1	1GE035	CF34-3A1	
CF34-8C1	5GE083	CF34-8C1	
CF34-8C5	6GE092	CF34-8C5	
CF6-45A2	3GE068	CF6-45A2	
CF6-50C	3GE070	CF6-50C	
CF6-50C1	3GE073	CF6-50C1	

CF6-50C2	3GE074	CF6-50C2	
CF6-50C2B	3GE078	CF6-50C2B	
CF6-50C2F	3GE078	CF6-50C2B	
CF6-50C2R	3GE072	CF6-50C2R	
CF6-50E1	3GE076	CF6-50E1	
CF6-50E2	3GE077	CF6-50E2	
CF6-6D	1GE001	CF6-6D	
CF6-6D1A	1GE002	CF6-6D1A	
CF6-6K	1GE003	CF6-6K	
CF6-80A	1GE010	CF6-80A	
CF6-80A2	1GE012	CF6-80A2	
CF6-80A3	1GE013	CF6-80A3	
CF6-80C2	2GE036	CF6-80C2A1	
CF6-80C2A1	2GE036	CF6-80C2A1	
CF6-80C2A2	2GE037	CF6-80C2A2	
CF6-80C2A3	2GE038	CF6-80C2A3	
CF6-80C2A5	2GE039	CF6-80C2A5	
CF6-80C2A5F	3GE056	CF6-80C2A5F	
CF6-80C2A8	2GE040	CF6-80C2A8	
CF6-80C2B	2GE041	CF6-80C2B1	
CF6-80C2B1	2GE041	CF6-80C2B1	
CF6-80C2B1F	2GE045	CF6-80C2B1F	
CF6-80C2B2	2GE042	CF6-80C2B2	
CF6-80C2B2F	2GE046	CF6-80C2B2F	
CF6-80C2B4	2GE043	CF6-80C2B4	
CF6-80C2B4F	2GE047	CF6-80C2B4F	
CF6-80C2B4FA	2GE047	CF6-80C2B4F	
CF6-80C2B5F	3GE057	CF6-80C2B5F	
CF6-80C2B6	2GE044	CF6-80C2B6	
CF6-80C2B6F	2GE048	CF6-80C2B6F	
CF6-80C2B7	2GE055	CF6-80C2B7F	
CF6-80C2B7F	2GE055	CF6-80C2B7F	
CF6-80C2B8	3GE058	CF6-80C2B8FA	
CF6-80C2B8F	3GE058	CF6-80C2B8FA	
CF6-80C2B8FG01	3GE058	CF6-80C2B8FA	
CF6-80C2D1F	2GE049	CF6-80C2D1F	
CF6-80E1A2	2GE051	CF6-80E1A2	
CF6-80E1A3	5GE085	CF6-80E1A3	
CF6-80E1A4	4GE081	CF6-80E1A4	
CF700-2C			CF700-2D
CF700-2D			CF700-2D
CF700-2D2			CF700-2D
CF700-2D-2			CF700-2D
CFE738-1-1B	7PW080	PW308C	
CFM56-2A-2	1CM001	CFM56-2A series	
CFM56-2C1	1CM003	CFM56-2-C5	
CFM56-2C3	1CM003	CFM56-2-C5	
CFM56-3A-2	1CM004	CFM56-3-B1	
CFM56-3B1	1CM004	CFM56-3-B1	
CFM56-3B2	1CM005	CFM56-3B-2	
CFM56-3C1	1CM007	CFM56-3C-1	
CFM56-5A1	1CM008	CFM56-5-A1	
CFM56-5A3	1CM009	CFM56-5A3	
CFM56-5A4	4CM035	CFM56-5A4	
CFM56-5A5	4CM036	CFM56-5A5	
CFM56-5B1	2CM012	CFM56-5B1	
CFM56-5B1/2	2CM016	CFM56-5B1/2	
CFM56-5B1/2P	3CM020	CFM56-5B1/2P	
CFM56-5B1\2	2CM016	CFM56-5B1/2	
CFM56-5B1\2P	3CM020	CFM56-5B1/2P	
CFM56-5B1\P	3CM023	CFM56-5B1/P	
CFM56-5B2	2CM013	CFM56-5B2	
CFM56-5B2\P	3CM024	CFM56-5B2/P	
CFM56-5B3	4CM038	CFM56-5B3/2P	
CFM56-5B3/2P	4CM038	CFM56-5B3/2P	
CFM56-5B3/P	3CM025	CFM56-5B3/P	
CFM56-5B3\2P	4CM038	CFM56-5B3/2P	
CFM56-5B3\P	3CM025	CFM56-5B3/P	
CFM56-5B4	2CM014	CFM56-5B4	
CFM56-5B4/2	2CM018	CFM56-5B4/2	
CFM56-5B4/2P	3CM021	CFM56-5B4/2P	
CFM56-5B4/P	3CM026	CFM56-5B4/P	
CFM56-5B4\2	2CM018	CFM56-5B4/2	
CFM56-5B4\2P	3CM021	CFM56-5B4/2P	
CFM56-5B4\P	3CM026	CFM56-5B4/P	
CFM56-5B5/P	3CM027	CFM56-5B5/P	
CFM56-5B5\P	3CM027	CFM56-5B5/P	
CFM56-5B6	2CM019	CFM56-5B6/2	
CFM56-5B6/2	2CM019	CFM56-5B6/2	
CFM56-5B6/2P	3CM022	CFM56-5B6/2P	
CFM56-5B6/P	3CM028	CFM56-5B6/P	
CFM56-5B6\2	2CM019	CFM56-5B6/2	
CFM56-5B6\2P	3CM022	CFM56-5B6/2P	
CFM56-5B6\P	3CM028	CFM56-5B6/P	
CFM56-5B7/P	6CM044	CFM56-5B7/P	
CFM56-5B7\P	6CM044	CFM56-5B7/P	
CFM56-5B8/P	6CM044	CFM56-5B8/P	
CFM56-5C/P	1CM010	CFM56-5C2	
CFM56-5C2	1CM010	CFM56-5C2	
CFM56-5C2/F	1CM010	CFM56-5C2	

CFM56-5C2/G	1CM010	CFM56-5C2	
CFM56-5C3	1CM011	CFM56-5C3	
CFM56-5C3/F	1CM011	CFM56-5C3	
CFM56-5C3\F	1CM011	CFM56-5C3	
CFM56-5C4	2CM015	CFM56-5C4	
CFM56-7B20	3CM030	CFM56-7B20	
CFM56-7B22	3CM031	CFM56-7B22	
CFM56-7B24	3CM032	CFM56-7B24	
CFM56-7B26	3CM033	CFM56-7B26	
CFM56-7B27	3CM034	CFM56-7B27	
CFM56-7B27B1	3CM034	CFM56-7B27	
CJ610-1			CJ610-6
CJ610-4			CJ610-6
CJ610-5			CJ610-6
CJ610-6			CJ610-6
CJ610-8A			CJ610-6
CJ-805-3	1PW001	JT3D-3B	
CONWAY42MK540	1CM003	CFM56-2-C5	
CONWAY43D-550	1CM003	CFM56-2-C5	
CT7-5A2			CT7-5
CT7-7A			CT7-5
CT7-7C			CT7-5
CT7-9A			CT7-5
CT7-9B			CT7-5
CT7-9C			CT7-5
CT7-9C3			CT7-5
D-18T	1AA001	D-30 (Il series)	
D-30	1AA001	D-30 (Il series)	
D-30-2	1AA001	D-30 (Il series)	
D-30-3	1AA001	D-30 (Il series)	
D-30KP	1AA002	D-30KP-2	
D-30KU	1AA003	D-30KU	
D-30KU-154-II	1AA004	D-30KU-154	
D-36	1ZM001	D-36	
D-36-2A	1ZM001	D-36	
D-36-3	1ZM001	D-36	
D-36-3A	1ZM001	D-36	
D-36-4A	1ZM001	D-36	
DART12MK301			RDa7
DART-529-8E			RDa7
DART529-8E			RDa7
DART-529-8X			RDa7
DART529-8X			RDa7
DART542-10			RDa7
DART542-10B			RDa7
DART542-10J/K			RDa7
DART542-10J\K			RDa7
DART542-10K			RDa7
DART542-4			RDa7
DART-542-4			RDa7
DART543-10			RDa7
DART6MK510			RDa7
DART6MK511-7E			RDa7
DART6MK514			RDa7
DART6MK514-7			RDa7
DART6MK517-7			RDa7
DART6MK529-7E			RDa7
DART7MK520			RDa7
DART7MK525			RDa7
DART7MK528-7E			RDa7
DART7MK531			RDa7
DART7MK532-7			RDa7
DART7MK532-7L			RDa7
DART7MK532-7R			RDa7
DART7MK532-9			RDa7
DART7MK533			RDa7
DART7MK533-2			RDa7
DART7MK534-2			RDa7
DART7MK535-2			RDa7
DART7MK535-7R			RDa7
DART7MK536-2			RDa7
DART7MK536-7R			RDa7
DART7MK552			RDa7
DART7MK552-7R			RDa7
DART8MK550-2			RDa7
E-2000-11			RDa7
ELAND504			501D22A
FJ44-1A	1PW035	JT15D-1 series	
FJ44-2A	1PW035	JT15D-1 series	
FJ44-2C	1PW035	JT15D-1 series	
G Q -30-MK 2			TIO-540-J2B2
GE90-90B	3GE060	GE90-90B	
GE90-92B	3GE061	GE90-92B	
GE90-94B	6GE091	GE90-94B	
IO-320-DIAD			IO-320-DIAD
IO-360-B			IO-360-B
IO-540-K1B5			TIO-540-J2B2
IO-540-K1C5			TIO-540-J2B2
J60-P-3			J57-P-22
JT12A-6	1PW035	JT15D-1 Series	

JT12A-6A	1PW035	JT15D-1 Series	
JT12A-8	1PW035	JT15D-1 Series	
JT12A-8A	1PW035	JT15D-1 Series	
JT12A-8N	1PW035	JT15D-1 Series	
JT15D-1A	1PW035	JT15D-1 series	
JT15D-1B	1PW035	JT15D-1 Series	
JT15D-4	1PW036	JT15D-4 series	
JT15D-4B	1PW036	JT15D-4 series	
JT15D-4D	1PW036	JT15D-4 series	
JT15D-5	1PW037	JT15D-5, -5A, -5B	
JT15D-5A	1PW037	JT15D-5, -5A, -5B	
JT15D-5D	1PW038	JT15D-5C	
JT3D-1	1PW001	JT3D-3B	
JT3D-3B	1PW001	JT3D-3B	
JT3D-7	1PW003	JT3D-7 series	
JT3D-7A	1PW003	JT3D-7 series	
JT4A-11	1PW001	JT3D-3B	
JT8D-11	1PW008	JT8D-11	
JT8D-11A	1PW008	JT8D-11	
JT8D-15	1PW010	JT8D-15	
JT8D-15A	1PW011	JT8D-15A	
JT8D-17	1PW013	JT8D-17	
JT8D-17A	1PW014	JT8D-17A	
JT8D-17R	1PW016	JT8D-17R	
JT8D-209	1PW017	JT8D-209	
JT8D-217	4PW068	JT8D-217	
JT8D-217A	4PW069	JT8D-217A	
JT8D-217C	4PW070	JT8D-217C	
JT8D-219	4PW071	JT8D-219	
JT8D-219C	4PW071	JT8D-219	
JT8D-7	1PW005	JT8D-7 series	
JT8D-7A	1PW005	JT8D-7 series	
JT8D-7B	1PW005	JT8D-7 series	
JT8D-9	1PW007	JT8D-9 series	
JT8D-9A	1PW007	JT8D-9 series	
JT9D-20	1PW031	JT9D-20	
JT9D-20J	1PW032	JT9D-20J	
JT9D-3A	1PW020	JT9D-7	
JT9D-59A	1PW033	JT9D-59A	
JT9D-7	1PW020	JT9D-7	
JT9D-70A	1PW034	JT9D-70A	
JT9D-7A	1PW021	JT9D-7A	
JT9D-7AH	1PW021	JT9D-7A	
JT9D-7ASP	1PW021	JT9D-7A	
JT9D-7AW	1PW021	JT9D-7A	
JT9D-7F	1PW023	JT9D-7F	
JT9D-7FW	1PW023	JT9D-7F	
JT9D-7J	1PW024	JT9D-7J	
JT9D-7Q	1PW025	JT9D-7Q	
JT9D-7Q3	1PW025	JT9D-7Q	
JT9D-7R4D	1PW026	JT9D-7R4D, -7R4D1	
JT9D-7R4D1	1PW026	JT9D-7R4D, -7R4D1	
JT9D-7R4E	1PW027	JT9D-7R4E, -7R4E1	
JT9D-7R4E1	1PW027	JT9D-7R4E, -7R4E1	
JT9D-7R4E4	1PW028	JT9D-7R4E4, -E1(H)	
JT9D-7R4G2	1PW029	JT9D-7R4G2	
JT9D-7R4H1	1PW030	JT9D-7R4H1	
JT9D-7W	1PW025	JT9D-7Q	
LF507-1F	1TL004	LF507-1F, -1H	
LF507-1H	1TL004	LF507-1F, -1H	
M45HMK 501	1RR001	M45H-01	
M-601A			PT6A-41
M-601D			PT6A-41
M-601E			PT6A-41
M-601F			PT6A-41
M-602			PT6A-41
NK-12MA			501D22A
NK-6	1KK002	NK-8-2U	
NK-8-2U	1KK002	NK-8-2U	
NK-8-4	1KK002	NK-8-2U	
NK-86	1KK003	NK-86	
O-540-E4C5			TIO-540-J2B2
OLYMPUS593			OLYMPUS-593-610
PS-90A	1AA005	PS-90A	
PS-90AT	1AA005	PS-90A	
PT6A-114			PT6A-114
PT6A-114A			PT6A-114
PT6A-20			PT6A-20
PT6A-27			PT6A-27
PT6A-27A			PT6A-27
PT6A-28			PT6A-28
PT6A-34			PT6A-34
PT6A-36			PT6A-36
PT6A-41			PT6A-41
PT6A-42			PT6A-42
PT6A-45A			PT6A-45
PT6A-45B			PT6A-45
PT6A-45R			PT6A-45R
PT6A-50			PT6A-50
PT6A-60			PT6A-60, -60A, -60AG

PT6A-60A			PT6A-60, -60A, -60AG
PT6A-65AR			PT6A-65AR
PT6A-65B			PT6A-65B
PT6A-65R			PT6A-65R
PT6A-67			PT6A-67B
PT6A-67D			PT6A-67D
PT6A-67R			PT6A-67D
PW115			PW118
PW118			PW118
PW118A			PW118A
PW118B			PW118B
PW119B			PW119-B
PW119C			PW119-C
PW120			PW120
PW120A			PW120A
PW121			PW121
PW121A			PW121
PW123			PW123
PW123B			PW123B
PW123BB			PW123B
PW123C			PW123C
PW123D			PW123D
PW123E			PW123E
PW124B			PW124-B
PW125B			PW125-B
PW126			PW126A
PW126A			PW126A
PW127			PW127-A
PW127B			PW127-A
PW127E			PW127E
PW127F			PW127-C,F,J
PW127G			PW127-C,F,J
PW127H			PW127-C,F,J
PW-127H			PW127E
PW-127J			PW127E
PW150A			PW127-C,F,J
PW2037	1PW039	PW2037	
PW2040	1PW040	PW2040	
PW2043	1PW040	PW2040	
PW305A	1TL004	LF507-1F, -1H	
PW305B	1TL004	LF507-1F, -1H	
PW306A	1TL004	LF507-1F, -1H	
PW306B	1TL004	LF507-1F, -1H	
PW308C	1TL004	LF507-1F, -1H	
PW4052	1PW042	PW4056	
PW4056	1PW042	PW4056	
PW4060	1PW043	PW4060	
PW4062	1PW043	PW4060	
PW4077	2PW061	PW4077	
PW4084	2PW062	PW4084	
PW4084D	3PW065	PW4084D	
PW4090	3PW066	PW4090	
PW4098	5PW076	PW4098	
PW4152	1PW045	PW4152	
PW4156A	1PW047	PW4156	
PW4158	1PW048	PW4158	
PW4164	1PW049	PW4164	
PW4164A	1PW049	PW4164	
PW4168	5PW075	PW4168	
PW4168A	4PW067	PW4168A	
PW4460	1PW052	PW4460	
PW4462	1PW052	PW4460	
PW530A	1PW036	JT15D-4 series	
PW535A	1PW037	JT15D-5, -5A, -5B	
PW545A	1PW037	JT15D-5, -5A, -5B	
R-2000-11			R-1820
R-2800-51			R-1820
R-2800-75			R-1820
R-2800-83AM4A			R-1820
R-2800-97			R-1820
R-2800-99W			R-1820
R-2800C			R-1820
R-2800-CA15			R-1820
R-2800-CA18			R-1820
R-2800-CB16			R-1820
R-2800-CB17			R-1820
R-3350-18DA1			R-1820
R-3350-18DA4			R-1820
R-3350-18EA1			R-1820
R-3350-34			R-1820
R-3350-75			R-1820
R-3350-91			R-1820
R-4360			R-1820
RB211-22B	1RR002	RB211-22B	
RB211-22B-02	1RR002	RB211-22B	
RB211-22N	1RR002	RB211-22B	
RB211-524B	1RR005	RB211-524B series	
RB211-524B2	1RR005	RB211-524B series	
RB211-524B202	1RR005	RB211-524B series	
RB211-524B2-02	1RR005	RB211-524B series	

RB211-524B4	1RR005	RB211-524B series	
RB211-524B402	1RR005	RB211-524B series	
RB211-524B4-02	1RR005	RB211-524B series	
RB211-524B4I	1RR005	RB211-524B series	
RB211-524BO2	1RR005	RB211-524B series	
RB211-524C2	1RR006	RB211-524C2	
RB211-524D4	1RR008	RB211-524D4	
RB211-524G	1RR010	RB211-524G	
RB211-524G\H-T	4RR037	RB211-524H-T	
RB211-524H	1RR011	RB211-524H	
RB211-524H2	1RR011	RB211-524H	
RB211-524HT	4RR037	RB211-524H-T	
RB211-535C	1RR012	RB211-535C	
RB211-535E4	5RR038	RB211-535E4	
RB211-535E4-B	5RR039	RB211-535E4B	
RB211-535E4-C	5RR039	RB211-535E4B	
SPEY-MK -511-8	1RR016	SPEY Mk511	
SPEYMK 511-8	1RR016	SPEY Mk511	
SPEYMK506	1RR016	SPEY Mk511	
SPEYMK506-14D	1RR016	SPEY Mk511	
SPEYMK511	1RR016	SPEY Mk511	
SPEYMK511-14	1RR016	SPEY Mk511	
SPEYMK511-14D	1RR016	SPEY Mk511	
SPEYMK511-14W	1RR016	SPEY Mk511	
SPEYMK512	1RR016	SPEY Mk511	
SPEYMK512-14D	1RR016	SPEY Mk511	
SPEYMK512-14DW	1RR016	SPEY Mk511	
SPEYMK512DW	1RR016	SPEY Mk511	
SPEYMK514-7W	1RR016	SPEY Mk511	
SPEYMK514DW	1RR016	SPEY Mk511	
SPEYMK555-15	1RR018	SPEY Mk555	
SPEYMK555-15H	1RR018	SPEY Mk555	
SPEYMK555-15P	1RR018	SPEY Mk555	
T56A-1			T56 series I
T56A-15			T56-A-15
T56A-16			T56-A-16
T56A-7			T56-A-7
T56A-7B			T56-A-7
T56A-9D			T56-A-9
TAY-MK -610-8	1RR019	TAY Mk611-8	
TAYMK 610-8	1RR019	TAY Mk611-8	
TAY-MK -611	1RR019	TAY Mk611-8	
TAY-MK -611-8	1RR019	TAY Mk611-8	
TAYMK 611-8	1RR019	TAY Mk611-8	
TAYMK 620-15	1RR020	TAY Mk620-15	
TAYMK 650-15	1RR021	TAY Mk650-15	
TAYMK 651-54	1RR021	TAY Mk650-15	
TF33-P-100	1PW001	JT3D-3B	
TFE731-20	1AS002	TFE731-3	
TFE731-20AR	1AS002	TFE731-3	
TFE731-20BR	1AS002	TFE731-3	
TFE731-20R	1AS002	TFE731-3	
TFE731-2-1C	1AS001	TFE731-2-2B	
TFE731-2-2B	1AS001	TFE731-2-2B	
TFE731-2-3B	1AS001	TFE731-2-2B	
TFE731-3-1C	1AS002	TFE731-3	
TFE731-3-1D	1AS002	TFE731-3	
TFE731-3-1E	1AS002	TFE731-3	
TFE731-3-1F	1AS002	TFE731-3	
TFE731-3-1G	1AS002	TFE731-3	
TFE731-3-1H	1AS002	TFE731-3	
TFE731-3-2B	1AS002	TFE731-3	
TFE731-3A-2B	1AS002	TFE731-3	
TFE731-3AR-2B	1AS002	TFE731-3	
TFE731-3B-100G	1AS002	TFE731-3	
TFE731-3B-100S	1AS002	TFE731-3	
TFE731-3C-100S	1AS002	TFE731-3	
TFE731-3C-200G	1AS002	TFE731-3	
TFE731-3CR-100S	1AS002	TFE731-3	
TFE731-3D-1C	1AS002	TFE731-3	
TFE731-3R-1D	1AS002	TFE731-3	
TFE731-3R-1H	1AS002	TFE731-3	
TFE731-40	1AS002	TFE731-3	
TFE731-40-1C	1AS002	TFE731-3	
TFE731-40R-200G	1AS002	TFE731-3	
TFE731-45R-2S	1AS002	TFE731-3	
TFE731-5AR-1C2	1AS002	TFE731-3	
TFE731-5AR-2C	1AS002	TFE731-3	
TFE731-5BR	1AS002	TFE731-3	
TFE731-5BR-1C	1AS002	TFE731-3	
TFE731-5BR-1H	1AS002	TFE731-3	
TFE731-5BR-2C	1AS002	TFE731-3	
TFE731-5R-1H	1AS002	TFE731-3	
TFE731-60	1AS002	TFE731-3	
TFE731-60-1C	1AS002	TFE731-3	
TPE331-10-252D			TPE331-10
TPE331-10-501C			TPE331-10
TPE331-10-511C			TPE331-10
TPE331-10GP511D			TPE331-10
TPE331-10GR-515H			TPE331-10

TPE331-10GT			TPE331-10
TPE331-10R-501C			TPE331-10
TPE331-10R-511C			TPE331-10
TPE331-10R-512C			TPE331-10
TPE331-10R-513C			TPE331-10
TPE331-10T-511D			TPE331-10
TPE331-10U-501G			TPE331-10
TPE331-10U-503G			TPE331-10
TPE331-10U-511G			TPE331-10
TPE331-10U-512G			TPE331-10
TPE331-10U-531G			TPE331-10
TPE331-10UA-511			TPE331-10
TPE331-10UA511G			TPE331-10
TPE331-10UA-511G			TPE331-10
TPE331-10UF-511			TPE331-10
TPE331-10UF-512			TPE331-10
TPE331-10UF-513			TPE331-10
TPE331-10UF513H			TPE331-10
TPE331-10UF-513H			TPE331-10
TPE331-10UF-516			TPE331-10
TPE331-10UG-513			TPE331-10
TPE331-10UG513H			TPE331-10
TPE331-10UG-514			TPE331-10
TPE331-10UG514H			TPE331-10
TPE331-10UG-514H			TPE331-10
TPE331-10UG-515			TPE331-10
TPE331-10UG-516			TPE331-10
TPE331-10UG516H			TPE331-10
TPE331-10UGR513			TPE331-10
TPE331-10UGR-513			TPE331-10
TPE331-10UGR514			TPE331-10
TPE331-10UGR515			TPE331-10
TPE331-10UGR-516H			TPE331-10
TPE331-10UR-513			TPE331-10
TPE331-10UR513H			TPE331-10
TPE331-10UR-513H			TPE331-10
TPE331-10UR515H			TPE331-10
TPE331-11U-601G			TPE331-10
TPE331-11U-611G			TPE331-10
TPE331-11U-611U			TPE331-10
TPE331-11U-612G			TPE331-10
TPE331-12JR701C			TPE331-12
TPE331-12U-701G			TPE331-12
TPE331-12UA-701			TPE331-12
TPE331-12UA701H			TPE331-12
TPE331-12UA702H			TPE331-12
TPE331-12UA-705			TPE331-12
TPE331-12UAR701			TPE331-12
TPE331-12UAR-701			TPE331-12
TPE331-12UAR-705H			TPE331-12
TPE331-12UH701G			TPE331-12
TPE331-12UH-701G			TPE331-12
TPE331-12UH-702			TPE331-12
TPE331-12UHR701			TPE331-12
TPE331-12UHR-701			TPE331-12
TPE331-12UHR-701G			TPE331-12
TPE331-12UHR-701H			TPE331-12
TPE331-14G-HR-802H			TPE331-14
TPE331-14G-HR-805H			TPE331-14
TPE331-14GR			TPE331-14
TPE331-14GR-801			TPE331-14
TPE331-14GR802H			TPE331-14
TPE331-14GR805H			TPE331-14
TPE331-14GR-805H			TPE331-14
TPE331-1UGR-513			TPE331-2
TPE331-2-201A			TPE331-2
TPE331-3U-303G			TPE331-3
TPE331-3U-304G			TPE331-3
TPE331-3UW-303G			TPE331-3
TPE331-3UW-304G			TPE331-3
TPE331-3UW-511G			TPE331-3
TPE331-5-251C			TPE331-3
TPE331-5-252D			TPE331-3
TPE331-5A-252D			TPE331-3
TRENT553	6RR041	Trent 556-61	
TRENT556	6RR041	Trent 556-61	
TRENT556-61	6RR041	Trent 556-61	
TRENT768-60	3RR029	Trent 768	
TRENT772-60	3RR030	Trent 772	
TRENT772B-60	3RR030	Trent 772	
TRENT884	2RR026	Trent 884	
TRENT890B	2RR027	Trent 892	
TRENT892	2RR027	Trent 892	
TRENT892-17	2RR027	Trent 892	
TRENT892B	2RR027	Trent 892	
TRENT892B-17	2RR027	Trent 892	
TRENT895	5RR040	Trent 895	
TV-3-117VMA			PW127-A
TV7-117C			PW127-C,F,J
TWD-10B			PT6A-65B

TYNE-515\10			TYNE
TYNE515\10			TYNE
TYNE515-101W			TYNE
V2500-A1	1IA001	V2500-A1	
V2522-A5	3IA006	V2522-A5	
V2524-A5	3IA007	V2524-A5	
V2525-D5	1IA002	V2525-D5	
V2527-A5	1IA003	V2527-A5	
V2527E-A5	1IA003	V2527-A5	
V2527M-A5	1IA003	V2527-A5	
V2528-D5	1IA004	V2528-D5	
V2530-A5	1IA005	V2530-A5	
V2533-A5	3IA008	V2533-A5	
VIPER-520	1AS002	TFE731-3	
VIPER-521	1AS002	TFE731-3	
VIPER-522	1AS002	TFE731-3	
VIPER522	1AS002	TFE731-3	
VIPER-601	1AS002	TFE731-3	
VIPER601	1AS002	TFE731-3	
VIPER-MK-526	1AS002	TFE731-3	
WASP-2SD13-G			R-1820
WASP2SD13-G			R-1820
WJ5A-1			PW127E

APPENDIX G: Cruise Altitude Dispersion Distributions

The following table shows the distributions of cruise altitudes developed from analyzing large sets of ETMS radar data. Descriptions of the fields are also provided.

Aircraft Cateory : J = Jets and T = Turboprops.
Starting Range: Lower end of the trip distance category.
Ending Range: Higher end of the trip distance category.
Fraction: Probability of the altitude.
Altitude: Cruise altitude.

Aircraft Category	Starting Range (nm)	Ending Range (nm)	Fraction	Altitude (ft)
J	50	100	0.0095	5000
J	50	100	0.0784	7000
J	50	100	0.0857	9000
J	50	100	0.1393	11000
J	50	100	0.1297	13000
J	50	100	0.1511	15000
J	50	100	0.0482	17000
J	50	100	0.0379	19000
J	50	100	0.0366	21000
J	50	100	0.0405	23000
J	50	100	0.0369	25000
J	50	100	0.044	27000
J	50	100	0.0355	29000
J	50	100	0.0419	31000
J	50	100	0.0503	33000
J	50	100	0.0347	35000
J	100	150	0.0466	13000
J	100	150	0.1122	15000
J	100	150	0.109	17000
J	100	150	0.1323	19000
J	100	150	0.1332	21000
J	100	150	0.0883	23000
J	100	150	0.0442	25000
J	100	150	0.0417	27000
J	100	150	0.0421	29000
J	100	150	0.0517	31000
J	100	150	0.0923	33000
J	100	150	0.0689	35000
J	100	150	0.0374	37000
J	150	200	0.0509	17000
J	150	200	0.1351	19000
J	150	200	0.2148	21000
J	150	200	0.1753	23000
J	150	200	0.0783	25000
J	150	200	0.0964	27000
J	150	200	0.044	29000
J	150	200	0.0549	31000
J	150	200	0.056	33000
J	150	200	0.0662	35000
J	150	200	0.0281	37000
J	200	250	0.1015	21000
J	200	250	0.1542	23000
J	200	250	0.1544	25000
J	200	250	0.2024	27000
J	200	250	0.1127	29000
J	200	250	0.0878	31000
J	200	250	0.074	33000
J	200	250	0.0735	35000
J	200	250	0.0394	37000
J	250	300	0.0615	23000
J	250	300	0.122	25000
J	250	300	0.2037	27000
J	250	300	0.1408	29000
J	250	300	0.164	31000
J	250	300	0.1408	33000
J	250	300	0.116	35000
J	250	300	0.0511	37000
J	300	400	0.0159	23000
J	300	400	0.0633	25000
J	300	400	0.1704	27000
J	300	400	0.1508	29000
J	300	400	0.1857	31000
J	300	400	0.1953	33000
J	300	400	0.1424	35000

J	300	400	0 0762	37000
J	400	500	0 0881	27000
J	400	500	0 1151	29000
J	400	500	0 2	31000
J	400	500	0 2592	33000
J	400	500	0 1852	35000
J	400	500	0 1118	37000
J	400	500	0 0405	39000
J	500	1000	0 0271	27000
J	500	1000	0 0474	29000
J	500	1000	0 1801	31000
J	500	1000	0 2791	33000
J	500	1000	0 2295	35000
J	500	1000	0 1622	37000
J	500	1000	0 0746	39000
J	1000	1500	0 1731	31000
J	1000	1500	0 2437	33000
J	1000	1500	0 2406	35000
J	1000	1500	0 2313	37000
J	1000	1500	0 1113	39000
J	1500	2500	0 0883	31000
J	1500	2500	0 1719	33000
J	1500	2500	0 2875	35000
J	1500	2500	0 3202	37000
J	1500	2500	0 1321	39000
J	2500	3500	0 0368	31000
J	2500	3500	0 1237	33000
J	2500	3500	0 2181	35000
J	2500	3500	0 2989	37000
J	2500	3500	0 3225	39000
J	3500	9999	0 128	31000
J	3500	9999	0 2792	33000
J	3500	9999	0 2829	35000
J	3500	9999	0 2255	37000
J	3500	9999	0 0844	39000
T	50	100	0 0876	4500
T	50	100	0 1275	5500
T	50	100	0 1237	6500
T	50	100	0 1367	7500
T	50	100	0 1208	8500
T	50	100	0 0829	9500
T	50	100	0 0958	10500
T	50	100	0 0719	11500
T	50	100	0 0553	12500
T	50	100	0 0347	13500
T	50	100	0 0336	14500
T	50	100	0 0293	15500
T	100	150	0 0689	7000
T	100	150	0 1009	9000
T	100	150	0 1256	11000
T	100	150	0 1474	13000
T	100	150	0 1894	15000
T	100	150	0 1039	17000
T	100	150	0 0746	19000
T	100	150	0 0498	21000
T	100	150	0 0412	23000
T	100	150	0 0158	25000
T	100	150	0 0158	27000
T	100	150	0 0131	29000
T	100	150	0 018	31000
T	100	150	0 0355	33000
T	150	200	0 0261	7000
T	150	200	0 0737	9000
T	150	200	0 0915	11000
T	150	200	0 1232	13000
T	150	200	0 2537	15000
T	150	200	0 1937	17000
T	150	200	0 1144	19000
T	150	200	0 0827	21000
T	150	200	0 041	23000
T	200	250	0 0214	11000
T	200	250	0 0383	13000
T	200	250	0 1197	15000
T	200	250	0 1181	17000
T	200	250	0 0884	19000
T	200	250	0 1187	21000
T	200	250	0 1268	23000
T	200	250	0 0944	25000
T	200	250	0 0905	27000
T	200	250	0 0555	29000
T	200	250	0 0427	31000
T	200	250	0 0413	33000
T	200	250	0 0443	35000
T	250	300	0 0144	9000
T	250	300	0 0398	11000
T	250	300	0 051	13000
T	250	300	0 1746	15000
T	250	300	0 1859	17000
T	250	300	0 1704	19000
T	250	300	0 1652	21000

T	250	300	0 1512	23000
T	250	300	0 0475	25000
T	300	400	0 0101	9000
T	300	400	0 0288	11000
T	300	400	0 0306	13000
T	300	400	0 1096	15000
T	300	400	0 1327	17000
T	300	400	0 1961	19000
T	300	400	0 1953	21000
T	300	400	0 1898	23000
T	300	400	0 1069	25000
T	400	500	0 0194	9000
T	400	500	0 0436	11000
T	400	500	0 0322	13000
T	400	500	0 0548	15000
T	400	500	0 0566	17000
T	400	500	0 1368	19000
T	400	500	0 15	21000
T	400	500	0 2129	23000
T	400	500	0 1844	25000
T	400	500	0 0644	27000
T	400	500	0 045	29000
T	500	9999	0 0468	9000
T	500	9999	0 0549	11000
T	500	9999	0 0467	13000
T	500	9999	0 0788	15000
T	500	9999	0 1078	17000
T	500	9999	0 1838	19000
T	500	9999	0 1356	21000
T	500	9999	0 1307	23000
T	500	9999	0 1278	25000
T	500	9999	0 0511	27000
T	500	9999	0 0359	29000

APPENDIX H: Sample Cruise Track Dispersion Distributions

The following tables shows a sample of the distributions of track offsets from the Great Circle developed from analyzing large sets of ETMS radar data. Descriptions of the fields are also provided.

Aircraft Cateory : J = Jets and T = Turboprops.
Starting Range: Lower end of the trip distance category.
Ending Range: Higher end of the trip distance category.
Percent Along Great Circle: Point along the Great Circle path defined as a percentage of distance along the Great Circle starting from the departure airport.
Fraction: Probability of the offset distance.
Offset Distance: Perpendicular offset from the Great Circle at the specified point along the Great Circle.

Aircraft Category	Starting Range (nm)	Ending Range (nm)	Percent Along Great Circle (%)	Fraction	Offset Distance (nm)
J	200	250	20	0 0366	0
J	200	250	20	0 033	1
J	200	250	20	0 0318	1
J	200	250	20	0 0301	2
J	200	250	20	0 0288	2
J	200	250	20	0 0274	3
J	200	250	20	0 0279	3
J	200	250	20	0 0291	4
J	200	250	20	0 0304	4
J	200	250	20	0 0312	5
J	200	250	20	0 0285	5
J	200	250	20	0 0263	6
J	200	250	20	0 0249	6
J	200	250	20	0 024	7
J	200	250	20	0 0235	7
J	200	250	20	0 0229	8
J	200	250	20	0 0225	8
J	200	250	20	0 0226	9
J	200	250	20	0 022	9
J	200	250	20	0 0225	10
J	200	250	20	0 0229	10
J	200	250	20	0 0221	11
J	200	250	20	0 022	11
J	200	250	20	0 0216	12
J	200	250	20	0 022	12
J	200	250	20	0 0215	13
J	200	250	20	0 0214	13
J	200	250	20	0 0195	14
J	200	250	20	0 0182	14
J	200	250	20	0 0165	15
J	200	250	20	0 0163	15
J	200	250	20	0 0148	16
J	200	250	20	0 0148	16
J	200	250	20	0 0138	17
J	200	250	20	0 0134	17
J	200	250	20	0 0126	18
J	200	250	20	0 0128	18
J	200	250	20	0 0131	19
J	200	250	20	0 0124	19
J	200	250	20	0 0113	20
J	200	250	20	0 0109	20
J	200	250	20	0 0103	21
J	200	250	20	0 0094	21
J	200	250	20	0 0092	22
J	200	250	20	0 0081	22
J	200	250	20	0 0075	23
J	200	250	20	0 0072	23
J	200	250	20	0 0065	24
J	200	250	20	0 0064	24
J	200	250	20	0 006	25
J	200	250	20	0 0057	25
J	200	250	20	0 0056	26
J	200	250	20	0 0046	26
J	200	250	20	0 0045	27
J	200	250	20	0 0038	27
J	200	250	20	0 0024	28
J	200	250	20	0 0006	28
J	200	250	30	0 032	0
J	200	250	30	0 0295	1
J	200	250	30	0 0287	1
J	200	250	30	0 0275	2

J	200	250	30	0 0265	2
J	200	250	30	0 0251	3
J	200	250	30	0 024	3
J	200	250	30	0 0232	4
J	200	250	30	0 0227	4
J	200	250	30	0 0221	5
J	200	250	30	0 0231	5
J	200	250	30	0 0233	6
J	200	250	30	0 0243	6
J	200	250	30	0 0236	7
J	200	250	30	0 0217	7
J	200	250	30	0 0205	8
J	200	250	30	0 019	8
J	200	250	30	0 0192	9
J	200	250	30	0 0188	9
J	200	250	30	0 0184	10
J	200	250	30	0 0174	10
J	200	250	30	0 0175	11
J	200	250	30	0 0175	11
J	200	250	30	0 0173	12
J	200	250	30	0 0166	12
J	200	250	30	0 0164	13
J	200	250	30	0 0161	13
J	200	250	30	0 016	14
J	200	250	30	0 0159	14
J	200	250	30	0 0156	15
J	200	250	30	0 0153	15
J	200	250	30	0 0148	16
J	200	250	30	0 0144	16
J	200	250	30	0 0142	17
J	200	250	30	0 0136	17
J	200	250	30	0 014	18
J	200	250	30	0 014	18
J	200	250	30	0 0136	19
J	200	250	30	0 0134	19
J	200	250	30	0 0126	20
J	200	250	30	0 0121	20
J	200	250	30	0 0115	21
J	200	250	30	0 0108	21
J	200	250	30	0 0111	22
J	200	250	30	0 011	22
J	200	250	30	0 0107	23
J	200	250	30	0 0103	23
J	200	250	30	0 0096	24
J	200	250	30	0 0093	24
J	200	250	30	0 0093	25
J	200	250	30	0 0089	25
J	200	250	30	0 0092	26
J	200	250	30	0 0085	26
J	200	250	30	0 0085	27
J	200	250	30	0 0077	27
J	200	250	30	0 0073	28
J	200	250	30	0 0074	28
J	200	250	30	0 007	29
J	200	250	30	0 0063	29
J	200	250	30	0 0057	30
J	200	250	30	0 0053	30
J	200	250	30	0 0046	31
J	200	250	30	0 0044	31
J	200	250	30	0 0042	32
J	200	250	30	0 0041	32
J	200	250	30	0 0041	33
J	200	250	30	0 0038	33
J	200	250	30	0 0036	34
J	200	250	30	0 001	34
J	200	250	40	0 0312	0
J	200	250	40	0 0295	1
J	200	250	40	0 0278	1
J	200	250	40	0 0261	2
J	200	250	40	0 0247	2
J	200	250	40	0 0235	3
J	200	250	40	0 0222	3
J	200	250	40	0 0212	4
J	200	250	40	0 0206	4
J	200	250	40	0 0209	5
J	200	250	40	0 02	5
J	200	250	40	0 0205	6
J	200	250	40	0 0205	6
J	200	250	40	0 0208	7
J	200	250	40	0 0209	7
J	200	250	40	0 0212	8
J	200	250	40	0 0201	8
J	200	250	40	0 0188	9
J	200	250	40	0 0173	9
J	200	250	40	0 016	10
J	200	250	40	0 0165	10
J	200	250	40	0 0157	11
J	200	250	40	0 0156	11
J	200	250	40	0 0154	12
J	200	250	40	0 015	12

J	200	250	40	0 0141	13
J	200	250	40	0 0136	13
J	200	250	40	0 0141	14
J	200	250	40	0 0135	14
J	200	250	40	0 0134	15
J	200	250	40	0 0132	15
J	200	250	40	0 0135	16
J	200	250	40	0 0135	16
J	200	250	40	0 0127	17
J	200	250	40	0 0128	17
J	200	250	40	0 0127	18
J	200	250	40	0 0127	18
J	200	250	40	0 0127	19
J	200	250	40	0 0125	19
J	200	250	40	0 0129	20
J	200	250	40	0 0127	20
J	200	250	40	0 0118	21
J	200	250	40	0 0119	21
J	200	250	40	0 0122	22
J	200	250	40	0 0117	22
J	200	250	40	0 0111	23
J	200	250	40	0 011	23
J	200	250	40	0 0104	24
J	200	250	40	0 0099	24
J	200	250	40	0 0101	25
J	200	250	40	0 0099	25
J	200	250	40	0 0097	26
J	200	250	40	0 0095	26
J	200	250	40	0 0092	27
J	200	250	40	0 009	27
J	200	250	40	0 0088	28
J	200	250	40	0 0088	28
J	200	250	40	0 0083	29
J	200	250	40	0 0079	29
J	200	250	40	0 0071	30
J	200	250	40	0 0068	30
J	200	250	40	0 0065	31
J	200	250	40	0 006	31
J	200	250	40	0 0055	32
J	200	250	40	0 0057	32
J	200	250	40	0 0055	33
J	200	250	40	0 0054	33
J	200	250	40	0 0055	34
J	200	250	40	0 0052	34
J	200	250	40	0 0052	35
J	200	250	40	0 0047	35
J	200	250	40	0 0046	36
J	200	250	40	0 0042	36
J	200	250	40	0 0026	37
J	200	250	40	0 0024	37
J	200	250	50	0 0325	0
J	200	250	50	0 0296	1
J	200	250	50	0 0266	1
J	200	250	50	0 0244	2
J	200	250	50	0 0234	2
J	200	250	50	0 0222	3
J	200	250	50	0 0211	3
J	200	250	50	0 0212	4
J	200	250	50	0 0212	4
J	200	250	50	0 0209	5
J	200	250	50	0 02	5
J	200	250	50	0 019	6
J	200	250	50	0 018	6
J	200	250	50	0 0175	7
J	200	250	50	0 0164	7
J	200	250	50	0 0171	8
J	200	250	50	0 0174	8
J	200	250	50	0 0186	9
J	200	250	50	0 0191	9
J	200	250	50	0 0204	10
J	200	250	50	0 0194	10
J	200	250	50	0 0177	11
J	200	250	50	0 0164	11
J	200	250	50	0 0155	12
J	200	250	50	0 0138	12
J	200	250	50	0 014	13
J	200	250	50	0 0136	13
J	200	250	50	0 0132	14
J	200	250	50	0 013	14
J	200	250	50	0 012	15
J	200	250	50	0 0123	15
J	200	250	50	0 012	16
J	200	250	50	0 0114	16
J	200	250	50	0 0114	17
J	200	250	50	0 0113	17
J	200	250	50	0 0108	18
J	200	250	50	0 0111	18
J	200	250	50	0 0115	19
J	200	250	50	0 0114	19
J	200	250	50	0 0116	20

J	200	250	50	0 0116	20
J	200	250	50	0 012	21
J	200	250	50	0 0129	21
J	200	250	50	0 0128	22
J	200	250	50	0 0127	22
J	200	250	50	0 0115	23
J	200	250	50	0 011	23
J	200	250	50	0 0102	24
J	200	250	50	0 0099	24
J	200	250	50	0 0094	25
J	200	250	50	0 0092	25
J	200	250	50	0 0087	26
J	200	250	50	0 0086	26
J	200	250	50	0 0084	27
J	200	250	50	0 0084	27
J	200	250	50	0 0088	28
J	200	250	50	0 0087	28
J	200	250	50	0 0084	29
J	200	250	50	0 0088	29
J	200	250	50	0 0086	30
J	200	250	50	0 0079	30
J	200	250	50	0 0078	31
J	200	250	50	0 0075	31
J	200	250	50	0 0073	32
J	200	250	50	0 0065	32
J	200	250	50	0 0069	33
J	200	250	50	0 0063	33
J	200	250	50	0 0061	34
J	200	250	50	0 0058	34
J	200	250	50	0 0056	35
J	200	250	50	0 0059	35
J	200	250	50	0 0057	36
J	200	250	50	0 0058	36
J	200	250	50	0 0053	37
J	200	250	50	0 0054	37
J	200	250	50	0 005	38
J	200	250	50	0 003	38
J	200	250	50	0 0024	39
J	200	250	60	0 0341	0
J	200	250	60	0 0305	1
J	200	250	60	0 0276	1
J	200	250	60	0 0265	2
J	200	250	60	0 0246	2
J	200	250	60	0 023	3
J	200	250	60	0 022	3
J	200	250	60	0 0214	4
J	200	250	60	0 021	4
J	200	250	60	0 0205	5
J	200	250	60	0 0201	5
J	200	250	60	0 0194	6
J	200	250	60	0 0187	6
J	200	250	60	0 0179	7
J	200	250	60	0 0176	7
J	200	250	60	0 0176	8
J	200	250	60	0 0166	8
J	200	250	60	0 017	9
J	200	250	60	0 0169	9
J	200	250	60	0 0172	10
J	200	250	60	0 0175	10
J	200	250	60	0 0175	11
J	200	250	60	0 0178	11
J	200	250	60	0 018	12
J	200	250	60	0 0173	12
J	200	250	60	0 0157	13
J	200	250	60	0 014	13
J	200	250	60	0 0135	14
J	200	250	60	0 0126	14
J	200	250	60	0 012	15
J	200	250	60	0 0114	15
J	200	250	60	0 0114	16
J	200	250	60	0 0108	16
J	200	250	60	0 0108	17
J	200	250	60	0 0107	17
J	200	250	60	0 0109	18
J	200	250	60	0 0115	18
J	200	250	60	0 011	19
J	200	250	60	0 0106	19
J	200	250	60	0 0108	20
J	200	250	60	0 0112	20
J	200	250	60	0 011	21
J	200	250	60	0 0111	21
J	200	250	60	0 0108	22
J	200	250	60	0 0109	22
J	200	250	60	0 011	23
J	200	250	60	0 011	23
J	200	250	60	0 0106	24
J	200	250	60	0 0112	24
J	200	250	60	0 0105	25
J	200	250	60	0 01	25
J	200	250	60	0 0104	26

J	200	250	60	0 0097	26
J	200	250	60	0 0092	27
J	200	250	60	0 0093	27
J	200	250	60	0 0087	28
J	200	250	60	0 0086	28
J	200	250	60	0 0085	29
J	200	250	60	0 0085	29
J	200	250	60	0 0085	30
J	200	250	60	0 0083	30
J	200	250	60	0 0083	31
J	200	250	60	0 0079	31
J	200	250	60	0 0076	32
J	200	250	60	0 0069	32
J	200	250	60	0 0063	33
J	200	250	60	0 0065	33
J	200	250	60	0 0062	34
J	200	250	60	0 0061	34
J	200	250	60	0 006	35
J	200	250	60	0 0059	35
J	200	250	60	0 0058	36
J	200	250	60	0 0055	36
J	200	250	60	0 0028	37
J	200	250	60	0 0025	37
J	200	250	60	0 0022	38
J	200	250	60	0 0022	38
J	200	250	60	0 0022	39
J	200	250	60	0 0005	39
J	200	250	70	0 0361	0
J	200	250	70	0 0329	1
J	200	250	70	0 029	1
J	200	250	70	0 0271	2
J	200	250	70	0 0257	2
J	200	250	70	0 0244	3
J	200	250	70	0 0232	3
J	200	250	70	0 0226	4
J	200	250	70	0 022	4
J	200	250	70	0 0213	5
J	200	250	70	0 0207	5
J	200	250	70	0 0205	6
J	200	250	70	0 0203	6
J	200	250	70	0 0203	7
J	200	250	70	0 0197	7
J	200	250	70	0 0191	8
J	200	250	70	0 017	8
J	200	250	70	0 0172	9
J	200	250	70	0 0166	9
J	200	250	70	0 0161	10
J	200	250	70	0 0151	10
J	200	250	70	0 0147	11
J	200	250	70	0 0151	11
J	200	250	70	0 0153	12
J	200	250	70	0 0156	12
J	200	250	70	0 0158	13
J	200	250	70	0 0156	13
J	200	250	70	0 0158	14
J	200	250	70	0 0154	14
J	200	250	70	0 0161	15
J	200	250	70	0 0156	15
J	200	250	70	0 0148	16
J	200	250	70	0 0137	16
J	200	250	70	0 0129	17
J	200	250	70	0 0126	17
J	200	250	70	0 0122	18
J	200	250	70	0 0123	18
J	200	250	70	0 0119	19
J	200	250	70	0 0113	19
J	200	250	70	0 0109	20
J	200	250	70	0 0107	20
J	200	250	70	0 0103	21
J	200	250	70	0 0107	21
J	200	250	70	0 0105	22
J	200	250	70	0 0104	22
J	200	250	70	0 0096	23
J	200	250	70	0 0098	23
J	200	250	70	0 0096	24
J	200	250	70	0 0093	24
J	200	250	70	0 009	25
J	200	250	70	0 009	25
J	200	250	70	0 009	26
J	200	250	70	0 0087	26
J	200	250	70	0 0094	27
J	200	250	70	0 0092	27
J	200	250	70	0 0086	28
J	200	250	70	0 0079	28
J	200	250	70	0 0082	29
J	200	250	70	0 008	29
J	200	250	70	0 0079	30
J	200	250	70	0 0075	30
J	200	250	70	0 0079	31
J	200	250	70	0 0078	31

J	200	250	70	0 0074	32
J	200	250	70	0 0068	32
J	200	250	70	0 0063	33
J	200	250	70	0 0062	33
J	200	250	70	0 0059	34
J	200	250	70	0 0056	34
J	200	250	70	0 0041	35
J	200	250	70	0 0027	35
J	200	250	70	0 0023	36
J	200	250	70	0 0022	36
J	200	250	70	0 0021	37
J	200	250	70	0 0015	37
J	200	250	80	0 0429	0
J	200	250	80	0 037	1
J	200	250	80	0 0322	1
J	200	250	80	0 0297	2
J	200	250	80	0 0275	2
J	200	250	80	0 0268	3
J	200	250	80	0 0261	3
J	200	250	80	0 0253	4
J	200	250	80	0 0247	4
J	200	250	80	0 0243	5
J	200	250	80	0 0244	5
J	200	250	80	0 0239	6
J	200	250	80	0 0215	6
J	200	250	80	0 0212	7
J	200	250	80	0 0197	7
J	200	250	80	0 0186	8
J	200	250	80	0 0182	8
J	200	250	80	0 0184	9
J	200	250	80	0 0177	9
J	200	250	80	0 0179	10
J	200	250	80	0 0173	10
J	200	250	80	0 0171	11
J	200	250	80	0 0163	11
J	200	250	80	0 0161	12
J	200	250	80	0 0169	12
J	200	250	80	0 0167	13
J	200	250	80	0 017	13
J	200	250	80	0 0169	14
J	200	250	80	0 0161	14
J	200	250	80	0 0152	15
J	200	250	80	0 0151	15
J	200	250	80	0 0164	16
J	200	250	80	0 0172	16
J	200	250	80	0 017	17
J	200	250	80	0 0163	17
J	200	250	80	0 0159	18
J	200	250	80	0 0147	18
J	200	250	80	0 0123	19
J	200	250	80	0 0102	19
J	200	250	80	0 0097	20
J	200	250	80	0 0091	20
J	200	250	80	0 0094	21
J	200	250	80	0 0092	21
J	200	250	80	0 0094	22
J	200	250	80	0 0096	22
J	200	250	80	0 0095	23
J	200	250	80	0 0099	23
J	200	250	80	0 0104	24
J	200	250	80	0 0104	24
J	200	250	80	0 0099	25
J	200	250	80	0 0097	25
J	200	250	80	0 0098	26
J	200	250	80	0 0099	26
J	200	250	80	0 0097	27
J	200	250	80	0 0088	27
J	200	250	80	0 0072	28
J	200	250	80	0 0065	28
J	200	250	80	0 0061	29
J	200	250	80	0 0062	29
J	200	250	80	0 0043	30
J	200	250	80	0 004	30
J	200	250	80	0 0037	31
J	200	250	80	0 0032	31
J	200	250	80	0 0026	32

APPENDIX I: WWLMINET

INTRODUCTION

WWLMINET (for World Wide LMI NETwork) treats a world-wide network of airports and en route sectors. The initial implementation includes 257 airports on six continents, as shown below:

WWLMINET was developed using many algorithms and routines found in LMINET, a capacity and delay queuing model of the United States. LMINET covers US terminal airspace, en route airspace, runways and taxiways, represented by a system of queues within those airspace structures. Because WWLMINET is expected to expand in coverage, the current version is the initial version, or version 1.0.

WWLMINET operates in single day increments. The program operates in epochs, which are one hour long. A day includes 24 epochs. Due to time zone changes around the world, the preference is to refer to time periods as epochs rather than any specific hour of the day.

PURPOSE AND FUNCTION

WWLMINET was built to forecast capacity and delay for a representative set of worldwide airports for any given day throughout the foreseeable future, to support the System for Assessing Aviation's Global Emissions (SAGE) model. Since detailed flight time data for all worldwide flights are difficult to obtain, models such as WWLMINET are used to estimate additional flight time due to congestion and delays.

WWLMINET takes as input a demand of current or future flights and cycles individual flights through the queuing network, which in version 1.0 consists of airport queues. In the initial 257 airports represented, flight demand is presented to the queues in accordance with a queuing network. If more aircraft demand service at a given queue than can be accommodated in a given epoch, the excess demand is delayed to the next epoch. The aircraft may be delayed but still accommodated in the next epoch; its delay is recorded in an output delay file. If an aircraft is delayed two entire epochs without being serviced by the relevant queue (i.e., waits two hours to taxi, take-off, or land), then it is canceled. Cancellations are also recorded in an output file.

Flights in the schedule that land or depart from an airport not in the 257 explicitly modeled are said to use an out-of-network airport. Delays at the out-of-network airports are not recorded, since capacities at those airports are not calculated (i.e., assumed no delays at those airports). Since many of these airports do not experience delays currently, their abstraction does not significantly impact the accuracy of worldwide results. However, flights with an origin or destination at an out-of-network airport that are canceled due to congestion at a network airport are still reported.

Delays and cancellations are recorded in output files and are used by the movements database of the SAGE model. Net delays between a baseline run and a policy or future run are added to stand-alone flight time, so that the additional emissions can be calculated where needed. Cancellations indicate where a flight would not have taken place, because hypothesized demand exceeds capacity by more than reasonable amounts. Key causes of cancellations in the model include weather-reduced capacity at airports, or high growth in demand that is not met by capacity increases.

Currently, just airport delays are modeled. Inclusion of airspace capacity constraint modeling will be investigated in a future version.

MODEL COVERAGE

The eventual goal in coverage is for WWLMINET to model all the world's commercial airports and airspaces. Version 1.0 represents the core capability of WWLMINET; the analysis engine and the highest-impact data. However, many data elements remain to be filled in, in subsequent releases.

Due to time constraints and the amount of effort required to obtain detailed capacity type information and data, efforts were focused on highest impact areas first. There are approximately 40,000 scheduled flights in the US daily. There are approximately 70,000 scheduled flights in the world daily, and 25% of those occur in Europe. Because of the dominance of the United States and Europe in air travel, the preponderance of modeled capacity in WWLMINET version 1.0 is also in the US and Europe. WWLMINET contains 102 US airports, representing 85 percent of US commercial airline travel; 122 European airports, representing approximately a quarter of worldwide commercial travel, and 33 other airports, all together representing three-quarters of world commercial travel as defined by the OAG. The airports included were selected on the basis of having the highest number of operations as defined by the year 2000 OAG and creating the most air travel delays in their country. Systematic delay information outside of the United States and Europe is not readily available. Airports outside US and Europe with highest delays were identified by asking national airline representatives to name their most delayed airports. The 257 networked airports are shown below:

Number	OAG Id	Longitude	Latitude	Country	Airport	ICAO ID /OAG ID
1	ABQ	-106 6093	35 0404	US	Albuquerque Intnatl Airport, NM	KABQ
2	ALB	-73 803	42 7481	US	Albany County Airport, NY	KALB
3	ANC	-149 995	61 1743	US	Ted Stevens Anchorage Intnatl, AK	KANC
4	ATL	-84 4269	33 6404	US	The William B Hartsfield Atlanta Intnatl Airport, GA	KATL
5	AUS	-97 7017	30 2986	US	The Robert Mueller Municipal Airport, TX	KAUS
6	BDL	-72 6832	41 9389	US	Bradley Locks, MA	KBDL
7	BFL	-119 0568	35 4336	US	Bakersfield/Meadows Field, CA	KBFL
8	BHM	-86 7535	33 5629	US	Birmingham Intnatl Airport, AL	KBHM
9	BNA	-86 6782	36 1245	US	Nashville Metropolitan Airport,TN	KBNA

10	BOI	-116 2228	43 5644	US	Boise Air Terminal Gowen Field, ID	KBOI
11	BOS	-71 0052	42 3643	US	General Edward Lawrence Logan Intnatl Airport, MA	KBOS
12	BTR	-91 1496	30 5332	US	Baton Rouge Metropolitan/Ryan Field, LA	KBTR
13	BUF	-78 7322	42 9405	US	Buffalo Niagara Intnatl Airport, NY	KBUF
14	BUR	-118 3585	34 2006	US	Burbank-Glendale Airport, CA	KBUR
15	BWI	-76 6682	39 1754	US	Baltimore-Washington Intnatl Airport, MD	KBWI
16	CHS	-80 0405	32 8986	US	Charleston AFB/Intnatl, SC	KCHS
17	CLE	-81 8494	41 4109	US	Hopkins Intnatl Airport, Cleveland, OH	KCLE
18	CLT	-80 9431	35 214	US	Douglas Airport, Charlotte, NC	KCLT
19	CMH	-82 8889	39 9962	US	Columbus Intnatl Airport, OH	KCMH
20	COS	-104 7003	38 8058	US	City of Colorado Springs Municipal Airport, CO	KCOS
21	CRP	-97 5012	27 7704	US	Corpus Christi Intnatl Airport, TX	KCRP
22	CVG	-84 6622	39 0461	US	Cincinnati-Northern KY Airport, OH	KCVG
23	DAB	-81 0576	29 1799	US	Daytona Beach Intnatl Airport, FL	KDAB
24	DAL	-96 8518	32 8471	US	Love Field, Dallas/Fort Worth, TX	KDAL
25	DAY	-84 2194	39 9024	US	Dayton Intnatl Airport, OH	KDAY
26	DCA	-77 0377	38 8521	US	Washington National Airport, D C	KDCA
27	DEN	-104 667	39 8584	US	Denver Intnatl Airport, CO	KDEN
28	DFW	-97 0372	32 896	US	Dallas-Fort Worth Intnatl Airport, TX	KDFW
29	DSM	-93 6607	41 5349	US	Des Moines Intnatl, IA	KDSM
30	DTW	-83 3488	42 2121	US	Detroit Metropolitan Wayne County Airport, MI	KDTW
31	ELP	-106 3778	31 8067	US	El Paso Intnatl Airport, TX	KELP
32	EUG	-123 2187	44 1233	US	Eugene/Mahlon Intnatl Airport, OR	KEUG
33	EWR	-74 1685	40 693	US	Newark Intnatl Airport, NJ	KEWR
34	FAT	-119 7181	36 7762	US	Fresno Air Terminal, CA	KFAT
35	FLL	-80 1527	26 0726	US	Ft Lauderdale-Hollywood Intnatl, FL	KFLL
36	FNT	-83 7435	42 9655	US	Fresno/Yosemite Intnatl Airport, CA	KFNT
37	GFK	-97 1761	47 9493	US	Grand Forks Intnatl Airport, ND	KGFK
38	GRR	-85 5228	42 8808	US	Grand Rapids/Gerald R Ford Intnatl Airport, MI	KGRR
39	GSO	-79 9373	36 0977	US	Greensboro-High Point Airport, NC	KGSO
40	HNL	-157 9224	21 31673	US	Honolulu Intnatl Airport, HI	KHNL
41	HOU	-95 2789	29 6454	US	William P Hobby Airport, Houston, TX	KHOU
42	HPN	-73 7076	41 067	US	Westchester County Airport, NY	KHPN
43	IAD	-77 4558	38 9447	US	Dulles Intnatl Airport, D C	KIAD
44	IAH	-95 3397	29 9805	US	Houston Intercontinental Airport, TX	KIAH
45	ICT	-97 433	37 65	US	Wichita/Mid-Continent Airport, KS	KICT
46	IND	-86 2944	39 7173	US	Indianapolis Intnatl Airport, IN	KIND
47	ISP	-73 1002	40 7952	US	MacArthur Field, Long Island, NY	KISP
48	JAX	-81 6878	30 4941	US	Jacksonville Intnatl Airport, FL	KJAX
49	JFK	-73 7789	40 6398	US	John F Kennedy Intnatl Airport, NY	KJFK
50	JNU	-134 5763	58 35496	US	Juneau Intnatl Airport, AK	KJNU
51	LAN	-84 5874	42 7787	US	Lansing/Capital City, MI	KLAN
52	LAS	-115 1521	36 0806	US	McCarran Intnatl Airport, Las Vegas, NV	KLAS
53	LAX	-118 4081	33 9425	US	Los Angeles Intnatl Airport, CA	KLAX
54	LGA	-73 8726	40 7772	US	La Guardia Airport, NY	KLGA
55	LGB	-118 1516	33 8177	US	Daugherty Field, Long Beach, CA	KLGB
56	LIT	-92 2246	34 729	US	Little Rock/Adams Field, AR	KLIT
57	MCI	-94 7139	39 2976	US	Kansas City Intnatl Airport, Kansas City, MO	KMCI
58	MCO	-81 316	28 4289	US	Orlando Intnatl Airport, FL	KMCO
59	MDW	-87 7524	41 786	US	Midway Airport, Chicago, IL	KMDW
60	MEM	-89 9768	35 0435	US	Memphis Intnatl Airport, TN	KMEM
61	MIA	-80 2903	25 7932	US	Miami Intnatl Airport, Miami, FL	KMIA
62	MKE	-87 897	42 9468	US	General Mitchell Field, Milwaukee, WI	KMKE
63	MLB	-80 6458	28 1028	US	Melbourne Intnatl Airport, FL	KMLB
64	MSN	-89 3375	43 1388	US	Madison/Dane County Regional Truax Field, WI	KMSN
65	MSP	-93 2169	44 8805	US	Minneapolis-Saint Paul Intnatl Airport, MN	KMSP
66	MSY	-90 2579	29 9935	US	New Orleans Intnatl Airport, LA	KMSY
67	OAK	-122 2207	37 7213	US	Oakland Intnatl Airport, CA	KOAK
68	OKC	-97 6007	35 3931	US	Oklahoma City/Will Rogers World Airport, OK	KOKC

69	OMA	-95 8942	41 3025	US	Omaha/Eppley Field, NE	KOMA
70	ONT	-117 6012	34 056	US	Ontario Intnatl Airport, Ontario, CA	KONT
71	ORD	-87 9045	41 9796	US	Chicago O' Hare Intnatl Airport, IL	KORD
72	ORF	-76 2012	36 8946	US	Norfolk Intnatl Airport, VA	KORF
73	PBI	-80 0956	26 6832	US	Palm Beach Intnatl Airport, FL	KPBI
74	PDX	-122 5975	45 5887	US	Portland Intnatl Airport, OR	KPDX
75	PHF	-76 493	37 1319	US	Newport News/Williamsburg Intnatl Airport, VA	KPHF
76	PHL	-75 245	39 8704	US	Philadelphia Intnatl Airport, PA	KPHL
77	PHX	-112 0095	33 4361	US	Phoenix Sky Harbor Intnatl Airport, AZ	KPHX
78	PIT	-80 2329	40 4915	US	Pittsburgh Intnatl Airport, PA	KPIT
79	PVD	-71 4267	41 7235	US	Providence/Theodore Francis Green State Airport, RI	KPVD
80	RDU	-78 7874	35 8776	US	Raleigh-Durham Airport, NC	KRDU
81	RIC	-77 3197	37 5052	US	Richmond Intnatl Airport, VA	KRIC
82	RNO	-119 7681	39 4986	US	Reno Cannon Intnatl Airport, NV	KRNO
83	ROC	-77 6724	43 1189	US	Rochester \Intnatl Airport, NY	KROC
84	RSW	-81 7552	26 5362	US	Southwest FL Intnatl Airport, Fort Myers, FL	KRSW
85	SAN	-117 1897	32 7336	US	Lindbergh Field, San Diego, CA	KSAN
86	SAT	-98 4698	29 5337	US	San Antonio Intnatl Airport,TX	KSAT
87	SBA	-119 8404	34 4262	US	Santa Barbara Municipal Airport, CA	KSBA
88	SDF	-85 7363	38 1741	US	Standiford Field, Louisville, KY	KSDF
89	SEA	-122 3093	47 449	US	Seattle-Tacoma Intnatl Airport, WA	KSEA
90	SFO	-122 3748	37 619	US	San Francisco Intnatl Airport, CA	KSFO
91	SJC	-121 929	37 3619	US	San Jose Intnatl Airport, CA	KSJC
92	SLC	-111 9778	40 7884	US	Salt Lake City Intnatl Airport, UT	KSLC
93	SMF	-121 5908	38 6954	US	Sacramento Metropolitan Airport, CA	KSMF
94	SNA	-117 8682	33 6757	US	John Wayne Airport, Orange County, CA	KSNA
95	STL	-90 36	38 7477	US	Lambert Field, Saint Louis, MO	KSTL
96	SWF	-74 1048	41 5041	US	Newburgh/Stewart Intnatl Airport, NY	KSWF
97	SYR	-76 1063	43 1112	US	Hancock Field, Syracuse, NY	KSYR
98	TPA	-82 5333	27 9755	US	Tampa/St Petersburg Airport, FL	KTPA
99	TUL	-95 8882	36 1984	US	Tulsa Intnatl Airport, OK	KTUL
100	TUS	-110 9413	32 1162	US	Tucson Intnatl Airport, AZ	KTUS
101	TVC	-85 5825	44 7408	US	Traverse City/Cherry Capital Airport, MI	KTVC
102	TYS	-83 9929	35 8125	US	McGhee Tyson Airport, Knoxville, TN	KTYS
103	ABZ	-2 19861	57 2042	UK	Aberdeen / Dyce	EGPD
104	ACE	-13 6036	28 9442	Spain	Lanzarote Arrecife	GCRR
105	AEP	-58 4058	-34 565	ARGENTINA	Buenos Aires -Newbery	SABE
106	AGP	-4 49778	36 6761	Spain	Malaga	LEMG
107	AJA	8 80361	41 925	France	Ajaccio Campo dell Oro	LFKJ
108	AKL	174 791	-37 01	NEW ZEALAND	AUCKLAND	NZAA
109	ALC	-0 558611	38 2836	Spain	Alicante	LEAL
110	AMS	4 76417	52 3081	Netherlands	Schiphol	EHAM
111	ARN	17 9219	59 6525	Sweden	Arlanda	ESSA
112	ATH	23 9444	37 9363	Greece	Athenai Hellinikon	LGAT
113	BCN	2 07972	41 2983	Spain	Barcelona	LEBL
114	BFS	-6 215	54 6575	UK	Belfast Int	EGAA
115	BGO	5 21972	60 2939	Norway	Bergen	ENBR
116	BHD	-5 8715	54 618	UK	Belfast City	EGAC
117	BHX	-1 74778	52 4539	UK	Birmingham	EGBB
118	BKK	100 608	13 9144	THAILAND	BANGKOK	VTBD
119	BLL	9 15306	55 7408	Denmark	Billund	EKBI
120	BLQ	11 2925	44 5306	Italy	Bologna	LIPE
121	BMA	17 9456	59 355	Sweden	Bromma (Stockholm)	ESSB
122	BNE	153 1175	-27 38416	AUSTRALIA	BRISBANE Queensland	YBBN
123	BOD	-0 713889	44 8294	France	Bordeaux Merignac	LFBD
124	BOG	-74 1422	4 70611	COLOMBIA	BOGOTA	SKBO
125	BOM	72 8667	19 0908	INDIA	BOMBAY	VABB
126	BRE	8 78778	53 0489	Germany	Bremen	EDDW
127	BRU	4 48583	50 9022	Belgium	Brussels Melsbroeck National	EBBR

128	BTS	17 2126	48 170167	Slovak	Stefanik Bratislava	LZIB
129	BUD	19 2633	47 4383	Hungary	Budapest Ferihegy	LHBP
130	CAN	113 265	23 1844	CHINA	GUANGZHOU Baiyum	ZGGG
131	CDG	2 54861	49 0097	France	Paris Charles De Gaulle	LFPG
132	CFE	3 16361	45 7867	France	Clermont Ferrand Auvergne	LFLC
133	CFU	19 9116	39 6019	Greece	Corfu Kerkyra, Ioannis Kapodistria	LGKR
134	CGH	-46 6558	-23 6256	BRAZIL	SAO PAULO-CONGONHAS	SBSP
135	CGN	7 14361	50 8672	Germany	Koln/Bonn	EDDK
136	CPH	12 6572	55 6186	Denmark	Copenhagen / Kastrup	EKCH
137	DRS	13 7672	51 1322	Germany	Dresden	EDDC
138	DUB	-6 25333	53 4311	Ireland	Dublin	EIDW
139	DUS	6 75806	51 2822	Germany	Dusseldorf	EDDL
140	EDI	-3 36139	55 9525	UK	Edinburgh	EGPH
141	EMA	-1 32444	52 8308	UK	East Midlands	EGNX
142	ERF	10 958	50 98	Germany	Erfurt	EDDE
143	ESB	32 9931	40 1244	Turkey	Ankara -Esenboga	LTAC
144	EZE	-58 5333	-34 8333	ARGENTINA	Buenos Aires-Pistarini	SAEZ
145	FAO	-7 96472	37 0128	Portugal	Faro	LPFR
146	FCO	12 2525	41 8111	Italy	Fiumicino (Rome)	LIRF
147	FLR	11 2031	43 8081	Italy	Florence Peretola, Vespucci	LIRQ
148	FNC	-16 7742	32 6906	Portugal	Funchal-Madeira	LPMA
149	FRA	8 57139	50 0344	Germany	Frankfurt am Main	EDDF
150	FUE	-13 8619	28 4508	Spain	Fuerteventura	GCFV
151	FUK	130 454	33 5811	JAPAN	FUKUOKA	RJFF
152	GCI	-2 60056	49 4361	UK	Guernsey	EGJB
153	GLA	-4 43194	55 8722	UK	Glasgow	EGPF
154	GOT	12 2936	57 6603	Sweden	Goteborg	ESGG
155	GRO	2 76333	41 9064	Spain	Gerona (Girona)	LEGE
156	GRU	-46 4667	-23 4333	BRAZIL	SAO PAULO-GUARULHOS	SBGR
157	GVA	6 11028	46 2397	Switzerland	Geneva	LSGG
158	HAJ	9 68507	52 461	Germany	Hannover	EDDV
159	HAM	9 98944	53 6319	Germany	Hamburg	EDDH
160	HEL	24 9664	60 3169	Finland	Helsinki	EFHK
161	HER	25 1775	35 3381	Greece	Heraklion, Iraklion, Nikos Kazanzakis	LGIR
162	HKG	114 2	22 3186	HONG KONG	HONG KONG	VHHH
163	HND	139 775	35 5475	JAPAN	TOKYO-HANEDA	RJTT
164	IBZ	1 37417	38 8742	Spain	Ibiza	LEIB
165	IST	28 8153	40 9767	Turkey	Ataturk Istanbul	LTBA
166	JER	-2 19444	49 2089	UK	Jersey	EGJJ
167	JNB	28 2428	-26 1339	SOUTH AFRICA	JOHANNESBURG	FAJS
168	JTR	25 4792	36 4014	Greece	Santorini	LGSR
169	KEF	-22 6058	63 9853	Iceland	Kefavik, Reykjavik	BIKF
170	KGS	27 0916	36 7957	Greece	Kos Hippocratis	LGKO
171	KHH	120 35	22 5754	TAIWAN	KADHSIUNG	RCKH
172	KIX	135 244	34 4273	JAPAN	KANSAI	RJBB
173	LBA	-1 65917	53 8658	UK	Leeds Bradford	EGNM
174	LCA	33 6303	34 8788	Cyprus	Larnaca	LCLK
175	LCY	0 054369	51 505256	UK	London City	EGLC
176	LEJ	12 2363	51 4239	Germany	Leipzig-Halle	EDDP
177	LGW	-0 188611	51 1478	UK	Gatwick (London)	EGKK
178	LHR	-0 459722	51 4769	UK	Heathrow (London)	EGLL
179	LIL	3 08861	50 5642	France	Lille-Lesquin	LFQQ
180	LIN	9 27861	45 4486	Italy	Milan Linate	LIML/LIMM
181	LIS	-9 13278	38 7728	Portugal	Lisbon	LPPT
182	LJU	14 4608	46 2247	Slovenia	Ljubljana	LJLJ
183	LPA	-15 3847	27 9303	Spain	Las Palmas Gran Canaria	GCLP
184	LTN	-0 366944	51 8742	UK	Luton (London)	EGGW
185	LUX	6 20556	49 6239	Luxembourg	Luxembourg	ELLX
186	LYS	5 08194	45 7269	France	Lyon Satolas, St Exupery	LFLL

187	MAD	-3 55944	40 4736	Spain	Madrid Barajas	LEMD
188	MAH	4 22	39 8631	Spain	Menorca	LEMH
189	MAN	-2 27472	53 3536	UK	Manchester	EGCC
190	MEL	144 8433	-37 6733	AUSTRALIA	MELBOURNE, VICTORIA	YMML
191	MEX	-99 0722	19 4353	MEXICO	MEXICO CITY	MMMX
192	MLA	14 4781	35 8586	Malta	Luqa	LMML
193	MLH	7 53	47 59	France	Bale-Mulhouse	LFSB
194	MMX	13 3561	55 5483	Sweden	Malmo-Sturup	ESMS/ESMM
195	MNL	121 019	14 50864	PHILIPPINES	MANILA: Ninoy Aquino	RPLL
196	MPL	3 9625	43 5844	France	Montpellier Mediterranee Langeudoc Roussillon	LFMT
197	MRS	5 21639	43 4369	France	Marseille Provence	LFML
198	MST	5 77694	50 9158	Netherlands	Maastricht-Aachen	EHBK
199	MUC	11 7875	48 3547	Germany	Munich Franz	EDDM
200	MXP	8 73083	45 6336	Italy	Milano Malpensa	LIMC
201	NAP	14 2889	40 8833	Italy	Naples, Capodichino	LIRN
202	NCE	7 21556	43 6656	France	Nice Cote d Azur	LFMN
203	NCL	-1 69	55 0372	UK	Newcastle	EGNT
204	NRT	140 391	35 7639	JAPAN	TOKYO-NARITA	RJAA
205	NTE	-1 60722	47 1572	France	Nantes Atlantique	LFRS
206	NUE	11 0792	49 4997	Germany	Nuremburg/Nurnburg	EDDN
207	OPO	-8 68138	41 248	Portugal	Porto; Francisco Sa Carneiro	LPPR
208	ORY	2 38139	48 7244	France	Paris Orly	LFPO
209	OSL	11 0856	60 2028	Norway	Oslo Gardermoen	ENGM
210	OTP	26 0867	44 575	Romania	Bucharest Otopeni	LROP
211	PEK	116 59	40 075	CHINA	BEIJING-CAPITAL	ZBAA
212	PMI	2 73056	39 5567	Spain	Palma de Mallorca	LEPA
213	PRG	14 2619	50 1017	Czech	Prague/Praha Ruzyne	LKPR
214	PXO	-16 3456	33 0669	Portugal	Funchal-Porto Santo	LPPS
215	RHO	28 0869	36 4056	Greece	Rhodos Diagoras	LGRP
216	RIX	23 9711	56 9236	Latvia	Riga	EVRA
217	RNS	-1 72944	48 0725	France	Rennes St Jacques	LFRN
218	RTM	4 43722	51 9569	Netherlands	Rotterdam	EHRD
219	SEL	126 4505	37 469	REPUBLIC OF KOREA	SEOUL: Inchon nee Kimpo, Gimpo	RKSI/RKSS
220	SHA	121 333	31 2	CHINA	SHANGHAI: Hongquio	ZSSS
221	SIN	103 991	1 35917	SINGAPORE	SINGAPORE: Changi	WSSS
222	SJU	-66 0018	18 4394	PUERTO RICO	SAN JUAN	TJSJ
223	SKG	22 9736	40 5194	Greece	Thessaloniki / Salonika, Makedonia	LGTS
224	SNN	-8 92083	52 7011	Ireland	Shannon	EINN
225	SOF	23 4075	42 6953	Bulgaria	Sofia	LBSF
226	SOU	-1 35528	50 9497	UK	Southampton	EGHI
227	STN	0 236667	51 8847	UK	Stansted (London)	EGSS
228	STR	9 21083	48 6886	Germany	Stuttgart	EDDS
229	SVG	5 63111	58 8811	Norway	Stavenger	ENZV
230	SVO	37 415	55 9717	RUSSIA	MOSCOW-SHEREMETYEVO	UUEE
231	SVQ	-5 8975	37 4192	Spain	Seville	LEZL
232	SXB	7 63278	48 5408	France	Strasbourg Entzheim	LFST
233	SXF	13 5225	52 38	Germany	Berlin Schonefeld	EDDB
234	SYD	151 1772	-33 9461	AUSTRALIA	SYDNEY N S W	YSSY
235	SZG	13 0039	47 795	Austria	Salzburg Mozart	LOWS
236	TFS	-16 5706	28 0428	Spain	Tenerife Sur Reina Sofia	GCTS
237	THF	13 4039	52 473	Germany	Berlin Tempelhof	EDDI
238	TLL	24 8328	59 4133	Estonia	Tallinn	EETN
239	TLS	1 38139	43 6239	France	Toulouse Blagnac	LFBO/LFBQ
240	TPE	121 232	25 080167	TAIWAN	CHIANG KAI SHEK INTL	RCTP
241	TRD	10 9258	63 4581	Norway	Trondheim Varnes	ENVA
242	TRN	7 65	45 2011	Italy	Turin Torino Caselle	LIMF
243	TSA	121 543	25 0714	TAIWAN	TAIPEI-SUNG SHAN	RCSS
244	TSF	12 1942	45 6478	Italy	Treviso San Angelico	LIPH
245	TXL	13 2877	52 5596	Germany	Berlin Tegel	EDDT

246	VCE	12 3519	45 5044	Italy	Venice	LIPZ/LIPV
247	VIE	16 5708	48 1108	Austria	Schwechat Vienna/Wien	LOWW
248	VNO	25 2858	54 6342	Lithuania	Vilnius	EYVI
249	VRN	10 8875	45 3953	Italy	Verona Villafranca	LIPX/LIPN
250	WAW	20 9689	52 1661	Poland	Warsaw	EPWA
251	YOW	-75 6692	45 3225	CANADA	OTTAWA MacDonald - Cartier Intl	CYOW
252	YQB	-71 3914	46 7933	CANADA	QUEBEC Jean Lesage	CYQB
253	YUL	-73 7414	45 4681	CANADA	MONTREAL	CYUL
254	YVR	-123 184	49 1939	CANADA	VANCOUVER B C	CYVR
255	YYC	-114 02	51 1139	CANADA	CALGARY	CYYC
256	ZAG	16 0742	45 7433	Croatia	Zagreb	LDZA
257	ZRH	8 54917	47 4594	Switzerland	Zurich	LSZH

INPUT DATA

WWLMINET requires the use of several input data files: (1) DepDmd, (2) OutArr, (3) WWSchedule, (4) CapData, and (5) Wx. The following sections describe each of data files and their contents.

DepDmd is the departure demand file for the 257 airports selected, accounting for the departures to all airports. It is used by WWLMINET to calculate hourly demand. It contains the following five fields:

- Airport Index (1-257)
- Airport code (IATA code)
- Epoch of departure time
- Demand in baseline year (whole number)
- Demand in forecast year (whole number)

The OutArr file contains the aggregate arrival demand to the network airports from out-of-network airports. It is used by WWLMINET to calculate hourly demand and contains following five fields:

- Airport Index (1-257)
- Airport code (IATA code)
- Epoch of departure time
- Demand in baseline year (whole number)
- Demand in forecast year (whole number)

WWschedule contains the unconstrained flight schedule (baseline or forecast) with "ticket" information. That is, each flight is listed individually as "tickets." Both the DepDmd and OutArr files were derived from the information contained within WWschedule which contains the following fields:

- Airport code
- Airport code
- Region
- Range
- Departure time in GMT
- Arrival time in GMT
- Carrier code in base year
- Flight number in base year
- Equipment type in base year
- Number of flights in base year

- Number of flights in forecast year (always 1 due to individual listings of multiple flights)
- Seat category in forecast year
- "Ticket" identifier in forecast year
- Departure airport index
- Epoch of departure time
- Arrival airport index (zero through 257)
- Epoch of arrival time (Epoch defined as GMT-5)

This file has the complete records of all flights, and each flight is listed individually. The flights are sorted by in-network, then departure airport, then departure epoch. This order is essential to WWLMINET. The network flights begin each file, followed by the flights with "8888" and then "9999" tickets. The out-of-network airports all share the index of "0."
Tickets are generated according to the following logic:

1. "Tickets" are generated for flights from networked airport to out-of-network airports (OON).
2. "8888" is assigned as the "ticket" value for a flight from OON to a networked airport.
3. "9999" is assigned as the "ticket" value for a flight from OON to OON.
4. The "ticket" is first assigned as 4 multiplied by the arrival airport index, and then add 1 if seat category is 1 or 2 if the seat category is 2 or above.

The CapData file contains airport capacity-related data with the following fields:

- WWLMINET airport number (1 -257)
- Weather category (0-4), where 0 is visual meteorological conditions (VMC), 1 is non-precision visibility, 2 is category 1 instrument meteorological conditions (IMC), 3 is category 2 IMC, and 4 is category 3 IMC.
- Number of points on Pareto optimal curve to be defined for that airport and weather category
- First point arrivals, also the number of arrivals corresponding to max departures (for that airport and weather category)
- First point departures, also maximum number of departures (for that airport and weather category)
- Second point arrivals, also the arrival rate corresponding to maximum number of arrivals (for that airport and weather category)
- Second point departures, also the number of departure corresponding to max arrivals (for that airport and weather category)
- Third point arrivals (for that airport and weather category) (optional)
- Third point departures (for that airport and weather category) (optional)

The Wx file is a weather data file containing airport-specific data that depict average weather conditions; the data depict a nice, sunny VMC day for the entire world. IMC weather files have not been provided, though they can easily be created using the format of the fields for the weather files:

- Epoch
- Ceiling in hundreds of feet
- Runway visual range in hundredths of a mile
- Windspeed and direction in the METAR format
- Temperature in Farenheit

OUTPUT DATA

WWLMINET outputs three files called Airport_Detail, Ticket_Detail, and Debug. Airport_Detail contains the per-aircraft delay information for the 257 modeled airports. En route delays are not modeled and so no en route delay is output.

The fields in the Airport_Detail file are shown below:

- Airport code
- Epoch
- Arrival demand
- Arrival runway delay in minutes
- Arrival taxi delay in minutes
- Departure demand
- Departure runway delay in minutes
- Departure taxi delay in minutes

This file provides the hourly delays for each of the 257 airports modeled in WWLMINET. The arrival taxi delay represents airport ground-based delay while the arrival runway delay represents the arrival airborne delay (e.g., during airborne holdings). For departure, just the taxi delay is used to model ground-based delay. Departure runway delay is just an artifact of the segregation of network queues and is not used.

The Ticket_Detail file contains the arrival fate of each in-network flight including cancellations. To discover if a flight listed in the WWSchedule file is canceled in WWLMINET as a result of delays, one must check the Ticket_Detail file which has the same order of flight listings as the WWSchedule file; thus to determine, for example, whether the tenth flight listed in WWSchedule is canceled, the tenth line of Ticket_Detail must be checked. The fields within Ticket_Detail are shown below:

- Departure airport code
- Arrival airport code
- Departure epoch
- Arrival epoch
- Status code, where 0 through 24 indicate the hour of arrival; 25 and 26 indicate cancellation of flight; 27 indicates an out of network arrival. Exact arrival times of flights arriving "out of network" cannot be known because the arrival airports' queues have not been modeled.

The Debug file is used to determine the causes of any errors that may have occurred, generally due to bad or inappropriate input data. Messages contained within this file help to determine the source of error.

THEORY AND MECHANICS OF WWLMINET

The queuing model is composed of airport models, a traffic model and a route model. Each of the airports in LMINET is represented by a set of linked arrival and demand queues. Additionally, each weather state that causes a change in configuration and capacity at each airport is represented by a new set of queuing variables. Each LMINET airport is modeled by an airport queuing network as shown below:

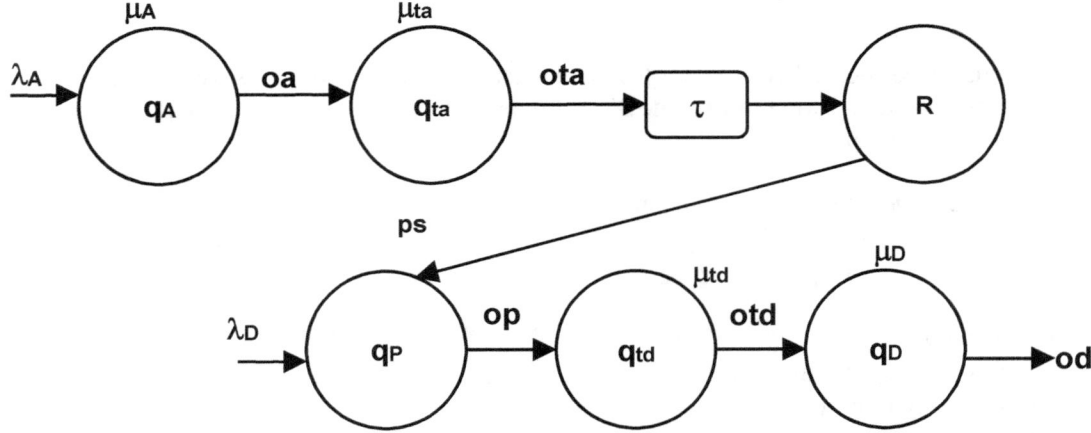

The capacity models begin with models of runway capacity. The basic concept of the runway capacity model is that controllers bring aircraft onto the common approach path and clear departures with sufficient time separation to insure that applicable regulations such as miles-in-trail, single-occupant rule, clearance between a departure and a subsequent arrival, etc. are met with specific levels of confidence. The confidence levels follow from considering means and standard deviations of aircraft operating parameters and controller-aircrew interactions in addition to certain airport characteristics.

As shown in the queing network fiture, traffic enters the arrival queue, q_A, according to an arrival process such as a Poisson distribution with parameter $\lambda_A(t)$. Upon service by the arrival server such as a Poisson service with parameter μ_{tA} and after the turnaround delay τ, arriving aircraft enter the ready-to-depart reservoir R. Each day's operations begin with a certain number of aircraft in this reservoir.

Departures require two services: an aircraft and a departure runway. Departures enter the queue for aircraft, q_P, according to the scheduled departure process. For example, this could be a Poisson process with parameter $\lambda_D(t)$. Departure aircraft are assigned by a process with service rate $\mu_P(t)$. When a departure aircraft is assigned, R is reduced by one. The departure leaves q_P and enters a taxi departure queue q_{td}, then a queue for a departure runway, q_D, where it is served according to the departure process characterized by $\mu_D(t)$.

Service at the queue for departing aircraft depends on the state of the ready-to-depart reservoir R. If R is not empty, then the service rate $\mu_P(t)$ is very large compared to one (service time is very short). If R is empty, then departing aircraft are supplied by output of the arrival queue, delayed by the turnaround time τ.

Queue lengths $q_A(t)$, $q_{tA}(t)$, $q_P(t)$, $q_{tD}(t)$, and $q_D(t)$ are integrated over time to give aircraft-minutes of delay for arrival, taxi-in, waiting-for-planes, taxi-out, and departure at each airport.
Airports for which no taxi delay information are available are estimated using regression analysis. The most significant predictor is existing hourly capacity.

The Pareto frontier is an isoquant that depicts the maximum possible arrivals and departures from an airport. More arrivals or departures can be accommodated if a particular runway handles only arrivals or only departures. The capacity of an airport overall depends on the number of runways it has, how it uses those runways for arrivals or departures or both, and the traffic interactions incident on operation of more than one runway at a time. An example Pareto frontier is shown in the following figure:

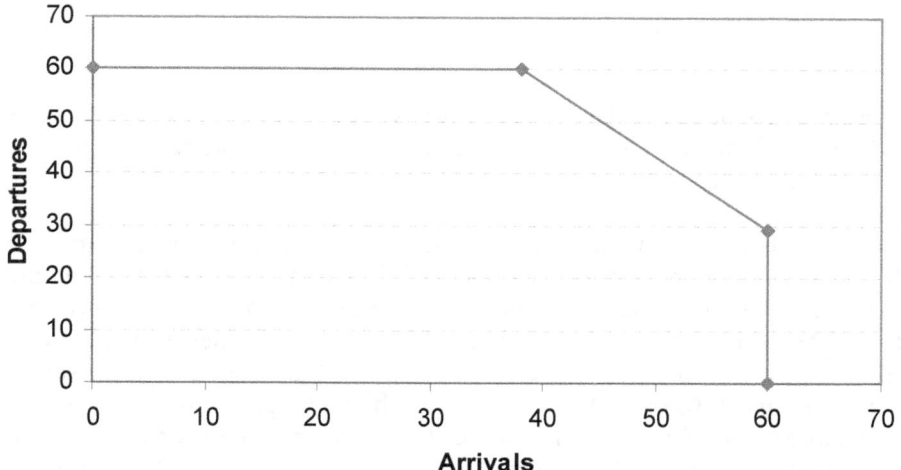

In WWLMINET, US airport queue settings were determined as a result of capacities modeled in the LMINET 2001 model. Capacities in LMINET 2001 are consistent with FAA Benchmarks. Capacities of European airports were modeled on the basis of runway engineering performance standards for Europe and held to be consistent with the *"European Database of Major Airports in the ECAC States"* 1998 Annual Report, published by Eurocontrol in December 1999. Capacities of non-US and non-European airports were modeled on the basis of runway engineering performance standards, that country's terminal and enroute procedures and separations, and confirmed against OAG flight mean hourly schedules for maximum capacities.

There are two kinds of aircraft in en route WWLMINET: 757s and 777s. Once an aircraft leaves the airport, it is assigned to a flying profile and speed consistent either with the 757 or the 777, whichever is closer to the aircraft's actual schedule type. The BADA flight profiles of these aircraft are used to derive the en route flying and thus determine the arrival time of the aircraft at the destination airport, where the aircraft re-enters a queuing network. The level of accuracy derived from using these aircraft types to mode aggregate travel times is sufficient in the epoch resolutions used within the model.

Taxi delays in WWLMINET are modeled parametrically based on a statistical relationship found between runway capacity and actual taxi delays, as reported by the BTS On-Time Performance data. This relationship was established based on US data. Actual taxi delay information from European and other non-US airports were obtained for Version 1.0 of WWLMINET. Based on the annual Eurocontrol ATM reports, a non-US taxi delay parameter was introduced into the estimation equations to tune the European and non-US airports taxi delays.

APPENDIX J: Uncertainty of Scaled SAGE Inventories using Sampled Datasets

OVERVIEW

A statistical anlysis was conducted to determine the best samples of the full SAGE movements data to use in forecasting and delay modeling. To simplify the assessment, a week was used as the base sampling unit in order to account for the effects of daily variations.

The analysis involved quantifying the sampling uncertainty of using a smaller set of movements data to represent a full year's worth of data. Instead of actual totals over all 52 weeks, they sampled totals are based on just the sampled weeks scaled-up to the full 52 weeks. Sampling uncertainty arises because the sampled weeks never exactly match the full 52 weeks. That is, they are never exactly proportional and therefore never scale-up to produce the same exact totals as the full 52 weeks. Some sampled weeks will match better, some will match worse. Moreover, the match will be generally better when a greater number of weeks (closer to the full 52) are included. The following sections quantify the sampling uncertainty expected depending upon how many weeks are sampled and how those weeks are chosen.

INPUT DATA AND ANALYSIS METHOD

In this analysis, sampling uncertainties were determined using the following full SAGE Version 1.0 daily inventories:

- Four years: 2000, 2001, 2002, and 2003.
- Six SAGE parameters: (1) Number of flights, (2) distance, (3) fuel burn, (4) CO, (5) HC, and (6) NOx. Since CO_2, H_2O, SOx are directly proportional to fuel burn, they are not included as one of the parameters (i.e., they would not provide any additional information).
- Eight world regions: (1) Africa, (2) Asia, (3) Australia and Oceania, (4) Eastern Europe, (5) Middle East, (6) North America and Caribbean, (7) South America, and (8) Western Europe and North Atlantic.

SAGE Version 1.0 data was necessarily used because Version 1.5 results cannot be generated until a specific sample has already been determined. That is, delays cannot be modeled in SAGE Version 1.5 until a sample has first been determined using Version 1.0 inventory data.

The Version 1.0 daily values were summed into weekly totals. The weekly total for any particular day is the sum over the seven-day period centered on that particular day. For example:

- Weekly total for 13 May: Sum over the seven days from 10 May through 16 May
- Weekly total for 14 May: Sum over the seven days from 11 May through 17 May

These weekly totals therefore overlap each other. Their purpose is to smooth out the day-to-day fluctuations in daily values. Had non-overlapping weeks been used instead, all seven-day periods would have to start on the same day of the week. This was not desired.

For any full SAGE variable, the true average weekly total is computed as follows:

$$\text{True average weekly total} = \text{True yearly total} / 52$$

If this true average weekly total is scaled up (multiplied by 52), it exactly yields the true yearly total—as is obvious from the above equation. For this reason, the best sampled weeks are those that best match this true average week. As a numerical measure of this match, the analysis used percentage offset from the true average week for each combination of year, region, and variable:

$$\text{\% Offset} = 100 \text{ (sampled week / true average week) / true average week}$$

The equation above represents an example for comparing just one week. Comparisons involving more than one week would be similarly accomplished.

Each week has 192 percentage offsets: 4 years multiplied by 8 world regions multiplied by 6 parameters. In general, each of these percentage offsets has a different value, depending upon how well that week matches the true average week for that particular year-region-parameter combination.

The best weeks are those with the smallest percentage offsets, as defined in two ways:

- Total range of possible offsets: A measure of the total width of a histogram of the range of offsets (or errors) from the 192 values. A small range corresponds to smaller inaccuracies for any combination of year-region-parameter.

- 95% uncertainty range: This numerical measure is related to a smooth bell-shaped curve that approximates a historgram of offsets. The curve is centered at the average value of all offsets and has a width that is proportional to the standard deviation of the offsets about this average value. Technically, the curve is the Gaussian approximation to the collection of offsets, and the 95% uncertainty range is the standard deviation of that Gaussian approximation divided by the square root of the number of weeks being sampled.

RESULTS

The analysis was conducted to determine the best one, two, and four weeks of samples. The following figures show histograms of offsets for each of these different samples.

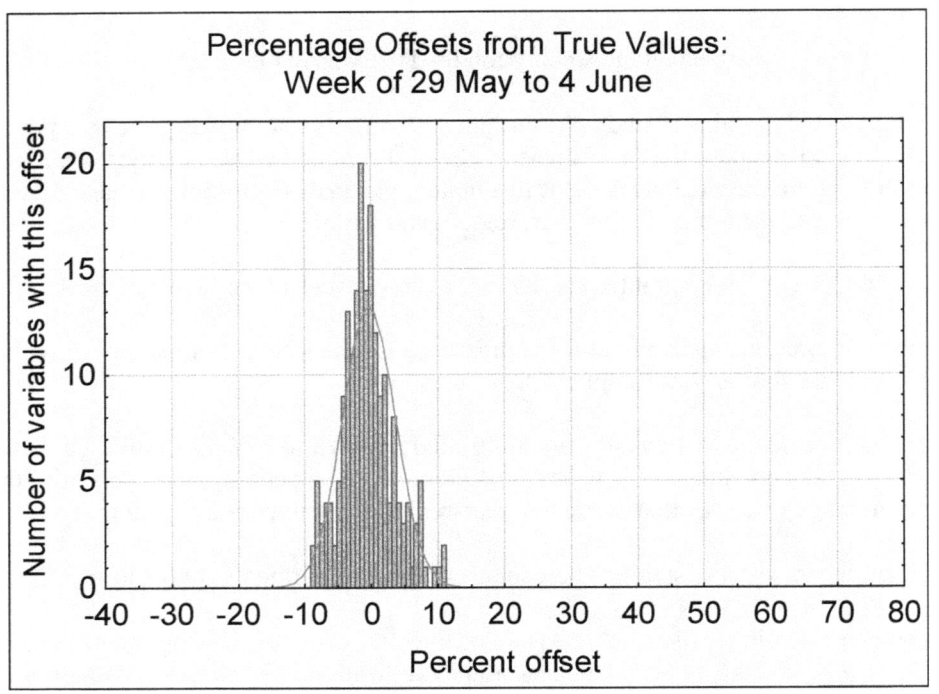

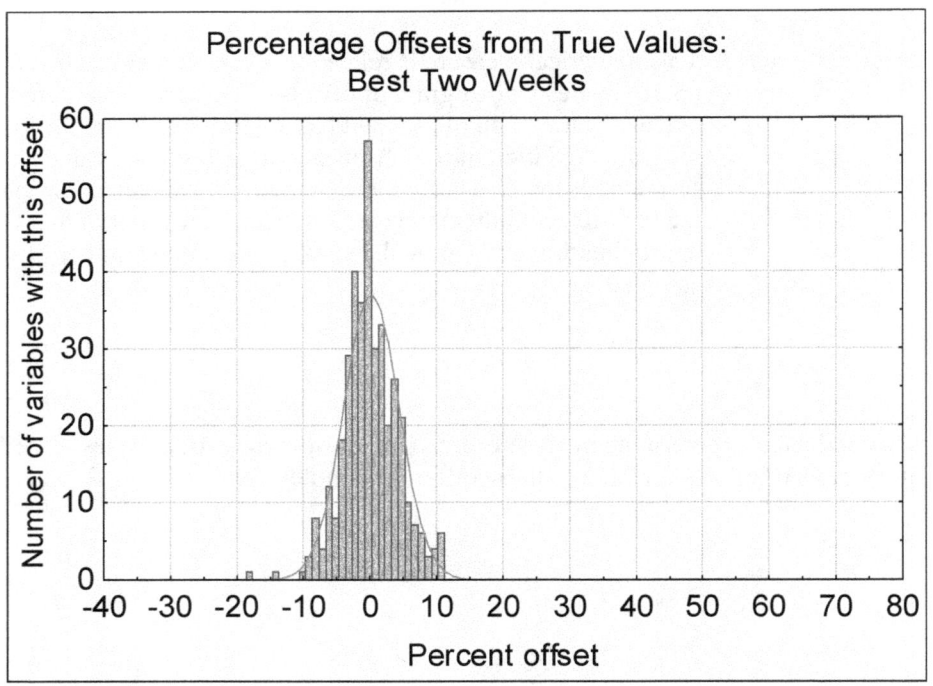

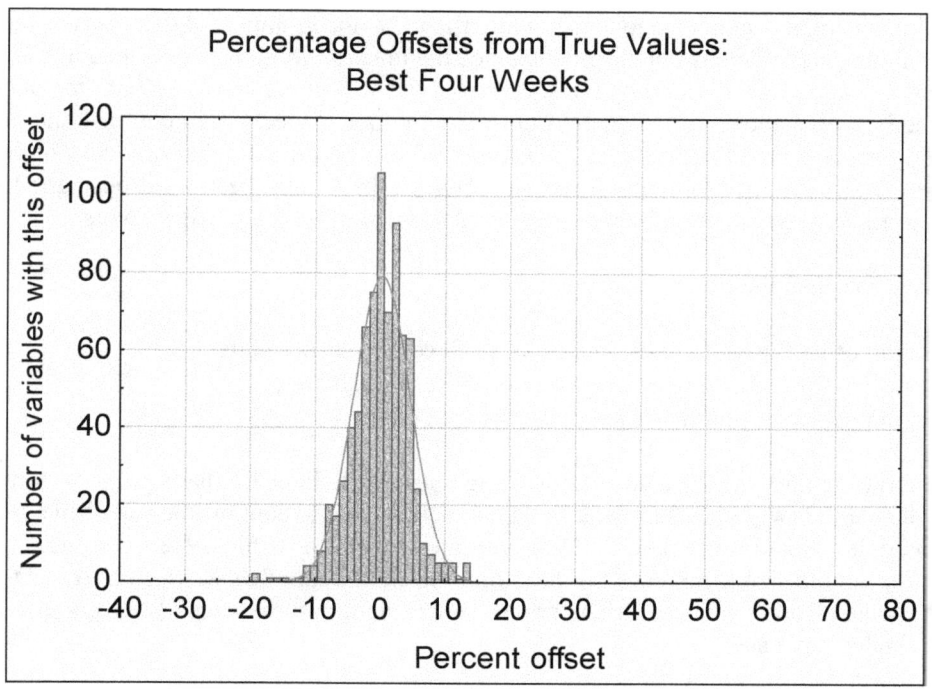

Based on the data derived from these histograms, the following table shows the corresponding statistics used to determine the best weeks.

Number of week(s)	Best weeks	Total range of possible offsets	95% uncertainty range		Uncertainty ratio: Best divided by Random (lower is better)
			Best week(s)	Random week(s)	
One	29 May to 4 Jun	−9% to +11%	±7.8%	±11.7%	0.7
Two	29 May to 4 Jun 25 Sep to 1 Oct	−18% to +11%	±4.0%	±8.5%	0.5
Four	29 May to 4 Jun 19 Oct to 25 Oct 18 Sep to 24 Sep 27 Sep to 3 Oct	−20% to +14%	±4.2%	±6.0%	0.7

As indicated in the table, the best weeks were chosen based on minimizing the range of possible offsets. The purpose in using this statistic was to reduce the effect of outliers (e.g., "special" weeks). The 95% uncertainty range is shown for reference purposes with comparisons to those for randomly selected weeks. By definition, the best week results will always be better than those for random weeks, and for the three sample sizes shown, the uncertainty ratios suggest reductions in uncertainties by approximately 30% to 50% when using best weeks as opposed to random weeks.

The effect of increasing the sample size is that while the 95% uncertainty range decreases, the range of possible offsets increases. This is not surprising since the uncertainty range is defined as the standard deviation of an average offset divided by the square root of the number of weeks (i.e., the uncertainty range is indirectly proportional to the square root of the number of weeks). In contrast, the range of offsets is simply based on the maximum and minimum offsets. As the number of weeks increases, slightly worse weeks are added to the total sample. That is, the second-best, third-best, and fourth-best weeks are progressively worse, thereby making the total range of possible offsets worse.

The criteria used to select the sample size are:

- Reduction of total range of possible offsets: fewer weeks are better.
- Reduction in 95% uncertainty ranges: more weeks are better.
- Reduction in model runtimes: fewer weeks are better.

With these criteria in mind, the 95% uncertainty ranges appear to indicate that a point of diminishing returns is reached at the two-week sample size. This would seem to support the use of two weeks for forecasting and delay modeling in SAGE. However, the range of possible offsets continues to increase noticeably as the sample size is increased. Therefore, the choice sample size was narrowed to either one or two weeks. In an effort to reduce outliers and to reduce runtimes, one week (29 May to 4 June) was chosen as the final sample size.

CONCLUSIONS

The statistical analysis did not show any clear indications for choosing a specific sample size. Therefore, based on a desire to minimize the effect of outliers and to reduce runtimes, a week (29 May to 4 June) was chosen as the final sample size for forecasting and delay modeling. Because four years' (2000-2003) worth of data were used in the analysis, the one week sample is not specific to any year.

The analysis was necessarily conducted using SAGE Version 1.0 inventory data because the sample size needed to be determined before Version 1.5 inventories can be developed. Similar assessments using Version 1.5 data will need to be conducted to help corroborate the sample choice. Finer data resolutions (i.e., finer than the year-region-parameter categories) and various sensitivity assessments can be conducted to further improve the analysis. Although forecasting can potentially be decoupled from delay modeling such that it could be based on a sample chosen from analyzing Version 1.5 data, the same one week (29 May to 4 June) is used for both forecasting and delay modeling for consistency.

APPENDIX K: Aircraft Performance Data

The following table provides a list of the BADA parameters used in SAGE.

BADA Parameter (SAGE Variable Name)	Units	Description
Model	N/A	BADA aircraft type
Type	N/A	Aircraft power source category (jet, turboprop, or piston)
MA_Ref	metric ton (t)	Reference mass
MA_Min	metric ton (t)	minimum mass
MA_Max	metric ton (t)	maximum mass
AD_Surf	m^2	Wing surface area
AD_Vstall_CR	knots	Stall speed (cruise)
AD_Cdo_CR	N/A	Parasitic drag coefficient (cruise)
AD_Cd2_CR	N/A	Induced drag coefficient (cruise)
AD_Vstall_IC	knots	Stall speed (initial climb)
AD_Cdo_IC	N/A	Parasitic drag coefficient (initial climb)
AD_Cd2_IC	N/A	Induced drag coefficient (initial climb)
AD_Vstall_TO	knots	Stall speed (takeoff)
AD_Cdo_TO	N/A	Parasitic drag coefficient (takeoff)
AD_Cd2_TO	N/A	Induced drag coefficient (takeoff)
AD_Vstall_AP	knots	Stall speed (takeoff)
AD_Cdo_AP	N/A	Parasitic drag coefficient (approach)
AD_Cd2_AP	N/A	Induced drag coefficient (approach)
AD_Vstall_LD	knots	Stall speed (landing)
AD_Cdo_LD	N/A	Parasitic drag coefficient (landing)
AD_Cd2_LD	N/A	Induced drag coefficient (landing)
AD_Cdo_LDG	N/A	Parasitic drag coefficient (landing gear)
ET_Ctc_1	N/A	1st maximum climb thrust coefficient
ET_Ctc_2	N/A	2nd maximum climb thrust coefficient
ET_Ctc_3	N/A	3rd maximum climb thrust coefficient
ET_Ctc_4	N/A	4th maximum climb thrust coefficient
ET_Ctc_5	N/A	5th maximum climb thrust coefficient
ET_Ctdes_Low	N/A	Low altitude thrust coefficient
ET_Ctdes_High	N/A	High altitude thrust coefficient
ET_Ctdes_Level	N/A	Level altitude thrust coefficient
ET_Ctdes_AP	N/A	Approach thrust coefficient
ET_Ctdes_LD	N/A	Landing thrust coefficient
FC_Thrust_1	N/A	1st fuel flow coefficient
FC_Thrust_2	N/A	2nd fuel flow coefficient
FC_Descent_1	N/A	1st descent fuel flow coefficient
FC_Descent_2	N/A	2nd descent fuel flow coefficient
FC_Cruise_Corr	N/A	Cruise fuel flow correction coefficient
Climb_Lowcas_LO	knots	1st standard climb VCAS (low)
Climb_Hicas_LO	knots	2nd standard climb VCAS (low)
Climb_Mc_LO	N/A	Standard climb Mach number above transition altitude (low)
Desc_Mc_LO	N/A	Standard descent Mach number above transition altitude (low)
Desc_Hicas_LO	knots	2nd standard descent VCAS (low)
Desc_Lowcas_LO	knots	1st standard descent VCAS (low)
Climb_Lowcas_AV	knots	1st standard climb VCAS (average)
Climb_Hicas_AV	knots	2nd standard climb VCAS (average)
Climb_Mc_AV	N/A	Standard climb Mach number above transition altitude (average)
Desc_Mc_AV	N/A	Standard descent Mach number above transition altitude (average)
Desc_Hicas_AV	knots	2nd standard descent VCAS (average)
Desc_Lowcas_AV	knots	1st standard descent VCAS (average)
Climb_Lowcas_HI	knots	1st standard climb VCAS (high)
Climb_Hicas_HI	knots	2nd standard climb VCAS (high)
Climb_Mc_HI	N/A	Standard climb Mach number above transition altitude (high)
Desc_Mc_HI	N/A	Standard descent Mach number above transition altitude (high)
Desc_Hicas_HI	knots	2nd standard descent VCAS (high)
Desc_Lowcas_HI	knots	1st standard descent VCAS (high)
Cruise_Mc_AV	N/A	Standard cruise Mach number

The following table provides a list of the INM parameters used in SAGE.

INM Parameter (SAGE Variable Name)	Units	Description
ACFT_ID	N/A	INM Aircraft type
AVG_COEFF_B	ft/lb	Takeoff distance coefficient
COEFF_E_TO	lbs	Corrected net thrust at zero speed (takeoff)
COEFF_E_CL	lbs	Corrected net thrust at zero speed (climb)
COEFF_F_TO	lbs/knot	Speed adjustment coefficient (takeoff)
COEFF_F_CL	lbs/knot	Speed adjustment coefficient (climb)
COEFF_GA_TO	bs/ft	Altitude adjustment coefficient (takeoff)
COEFF_GA_CL	bs/ft	Altitude adjustment coefficient (climb)
COEFF_GB_TO	lbs/ft2	Altitude-squared adjustment coefficient (takeoff)
COEFF_GB_CL	lbs/ft2	Altitude-squared adjustment coefficient (climb)
COEFF_H_TO	lbs/K	Temperature adjustment coefficient (takeoff)
COEFF_H_CL	lbs/K	Temperature adjustment coefficient (climb)
DIST_GR1	ft	1st takeoff ground roll distance for standard conditions
VGR1	knots	1st takeoff ground roll speed for standard conditions
THRUST_GR1	lbs	1st takeoff ground roll thrust for standard conditions
DIST_GR2	ft	2nd takeoff ground roll distance for standard conditions
VGR2	knots	2nd takeoff ground roll speed for standard conditions
THRUST_GR2	lbs	2nd takeoff ground roll thrust for standard conditions
DIST_AP1	ft	1st approach ground roll distance for standard conditions
VAP1	knots	1st approach ground roll speed for standard conditions
THRUST_AP1	lbs	1st approach ground roll thrust for standard conditions
DIST_AP2	ft	2nd approach ground roll distance for standard conditions
VAP2	knots	2nd approach ground roll speed for standard conditions
THRUST_AP2	lbs	2nd approach ground roll thrust for standard conditions
DIST_AP3	ft	3rd approach ground roll distance for standard conditions
VAP3	knots	3rd approach ground roll speed for standard conditions
THRUST_AP3	lbs	3rd approach ground roll thrust for standard conditions
NUM_ENGINES	N/A	Number of engines on aircraft

APPENDIX L: A Methodology to Account for Unscheduled and Canceled Commercial Flights

This appendix describes the work that was conducted in order to develop regression models that can be used to predict the number of unscheduled and canceled flights by airport. As this material is expected to be published in a journal article, its format and content has been left intact. Therefore, some of the material will overlap with other parts of this Technical Manual. Once the material is published, it will no longer reside in this appendix, and will instead be referenced through the journal.

1 Introduction

The Federal Aviation Administration's Office of Environment and Energy (FAA/AEE) with support from John A. Volpe National Transportation Systems Center (Volpe), the Massachusetts Institute of Technology (MIT) and the Logistics Management Institute (LMI) are developing the System for assessing Aviation's Global Emissions (SAGE) [FAA, 2005b][FAA, 2005c]. The development team envisions SAGE as an internationally accepted computer model that can be used for predicting and evaluating the effects of different policy and technology scenarios on aviation-related fuel burn, emissions, costs, aircraft performance, and industry responses. With regard to scope, the model is capable of analyses on a flight, airport, regional, and global level. This paper addresses one specific issue with regard to refining the SAGE model, namely the difference between scheduled and actual flight data. Presented herein is a methodology to account for unscheduled and canceled commercial aviation flights, including freight or cargo flights.

1.1 Statement of Need

Any individual or organization operating a flight (known as a carrier) provides scheduling information to the FAA and others sometime prior to each flight, regardless of the size of the aircraft or the nature of the carrier. The scheduling information is basic information on the flight operation, and usually contains the departure/arrival time, origin/destination airport, carrier name, and the type of aircraft. The scheduling information from various carriers is combined in a data set known as the Official Airline Guide (OAG) [FAA, 2005a], and can be used for a variety of different data analyses. However, recognizing that the OAG is based solely on scheduling information, and not on the actual operation of flights, two problems can arise. First, a carrier may operate a flight providing no prior scheduling information; and second, a carrier may cancel a scheduled flight for any number of reasons.

While the OAG is considered a close approximation of the number of flight operations, some analyses require a more precise measure of the number of flights. One example where higher precision is required is SAGE. For the SAGE model scheduling data is obtained, including origin/destination airport and information on the type of aircraft flown. For each scheduled flight SAGE computes the fuel burn and the emissions, which are dependent upon aircraft type and engine type. Since OAG is incomplete, SAGE uses other sources of flight information. One such source of actual flight data is the Enhanced Traffic Management System (ETMS) [Volpe, 2003], which is a compilation of flight plans and radar data covering all of North America and the United Kingdom.

Since ETMS geographic coverage is limited and because ETMS archives are either incomplete or non-existent prior to 2000, SAGE augments the ETMS flight plan data with data from OAG. For regions in which OAG is the primary source of flight plan data a methodology is needed to account for canceled and unscheduled flights. It has been shown that not accounting for canceled and unscheduled flights can introduce errors in SAGE fuel burn and emissions of 10% or more [Lee, 2005]. While the methodology described in this paper has been developed specifically for SAGE, its applicability is by no means limited to that model.

1.2 Objective and Scope

The objective of this paper is to present a comprehensive assessment of the number of canceled and unscheduled commercial aviation flights, and to develop a methodology to adjust the OAG flight plan data set accordingly. Comparisons of ETMS and OAG data are made for 2000, the first half of 2001 and for 2003 for flights departing from the United States only, regardless of destination. In developing the methodology, ETMS data is used as the "gold standard" for comparison with OAG data. For example, if a flight appears only in ETMS, it can be considered an unscheduled flight, and conversely, if a flight appears only in the OAG, it can be considered canceled. Once the number of unscheduled/canceled flights is quantified, the OAG data can be adjusted or calibrated to better represent actual flight data.

1.3 Document Outline

Section 2 will discuss the logic, parameters and assumptions for comparing the ETMS and the OAG data sets. Section 3 will discuss the overall results of comparing the two data sets; while Section 4 and 5 will deal specifically with a recommended methodology to account for unscheduled and canceled flights, respectively. Section 6 presents a validation of the recommended methodology using several independent data sets.

2 Data Processing

This section discusses how the two data sets, ETMS and OAG, were compared to one another. The purpose of matching up the same flight in ETMS and OAG is to identify flights as scheduled and flown, unscheduled, or canceled. If an individual flight is found in both ETMS and OAG, then it is assumed to be scheduled and flown. If a flight appears only in the ETMS data set, then it will be considered unscheduled, as no prior scheduling information was provided to the FAA, but the flight took place. And if a flight appears only in the OAG but not in the ETMS data set, then it will be considered canceled, as scheduling information was provided but no flight operation took place.

2.1 Description of OAG and ETMS Data Sets

As stated previously, OAG is scheduling information provided by the carrier and includes the following information: flight ID, carrier name, departure date/time, origin/destination airport, and aircraft type. ETMS is a compilation of flight plan and radar data that contains the following information: flight ID, departure/arrival date/time, origin/destination airport, user class, weight class, physical class, and aircraft type. The ETMS user class has five categories: commercial, freight, taxi, military, and general aviation. Commercial is any flight that is operating for commercial purposes (mostly passenger), freight is any flight that is conducted for the purpose of shipping cargo, taxi flights are similar to commercial flights, but are generally shorter range, military are flights conducted by any military establishment, and general aviation are personal civilian flights. The ETMS weight class is a parameter to help determine the physical mass of the aircraft and has three categories: small, large, and heavy. The small weight class includes everything from small, single-engine general aviation airplanes to 10 to 12 passenger jet aircraft. The large weight class includes the mid-range commercial aircraft, such as the Boeing 737 and the Airbus

A320 series aircraft, and also larger propeller driven aircraft like the C-130. The Heavy weight class includes aircraft such as the Boeing 767 and the Airbus A300 series aircraft. The physical class in ETMS is determined by the propulsion system, and has three categories: jet, prop, and turbo-prop.

2.2 Data Selection and Filtering

Five weeks of data were selected for this study: May 13th to 19th 2000, October 13th to 19th 2000, May 13th to 19th 2001, May 13th to 19th 2003, and October 13th to 19th 2003. The specific weeks were selected with the idea of representing seasonal variability, over a range of more recent years, while avoiding late 2001 and all of 2002 for obvious reasons. A few general filters were placed on all the data. First, the analysis was limited to a geographic region in which ETMS coverage of commercial flights is complete. Therefore the analysis included all commercial flights departing from a US airport (including Alaska and Hawaii), regardless of destination. The logic behind this was if the aircraft was departing from the US, it would appear in ETMS, and its destination would have no bearing on whether it would appear. The other general restriction was all flights with the user class of general aviation or military were excluded from the analysis.

One difficulty encountered in conducting this analysis is that a flight may on occasion be repeated in the ETMS data set; i.e., two separate flights in ETMS could be listed but contain the same flight ID and departure time/airport. Any duplicate flights in ETMS were deleted prior to analysis.

2.3 Data Comparison

In order to verify that a particular flight is in both the OAG and the ETMS data set, four flight parameters common to both OAG and ETMS were compared: flight ID, departure date, departure airport, and arrival airport. If all four values were identical in both ETMS and OAG, then it was considered the same flight. However, because the OAG and ETMS data sets are unique, each with its own format and ways of collecting data, a few methodological adjustments were required.

There is a fundamental difference in the flight IDs' in ETMS and OAG that requires attention. The flight ID is generally a three-character sequence (known as the carrier code) followed by a 3 to 5 digit numeric. In some instances, the OAG carrier code is only a two-character sequence. Consequently, the two-character OAG code needed to be mapped to the corresponding three-character ETMS code. In addition, some flights are flown under what is called a sub-carrier; a smaller carrier owned and operated by a larger carrier. For instance, Exp-Mesa, Exp-Piedmont, and Exp-Allegheny all fly under US Airways. This creates a disconnect when comparing the ETMS and OAG flight IDs. In the previous example, OAG will use the main carrier code (US Airways), while ETMS will use the sub-carrier code (Exp-Piedmont). Fortunately, OAG contains a longer description of the carrier, which also indicates the sub-carrier. So, for each OAG flight, the three-letter carrier code was mapped to the three-letter sub-carrier code. Finally, in some instances, the 3 to 5 digit number in the OAG flight ID would be preceded by additional digits that did not appear in the ETMS flight ID; these digits were removed for the analysis.

3 Results of Comparative Analysis

Table 3.1 presents a comparison of the OAG to the ETMS data sets for the five weeks included in the analysis. "Flown and Scheduled" indicates the combined number of flights located in both the ETMS and OAG data sets, "Unscheduled" indicates the number of flights in the ETMS data set, and "Canceled" indicates the number of flights located in the OAG data set.

Table 3.1. Results of ETMS/OAG Comparison

Date	Flown and Scheduled	Unscheduled	Canceled	% Unscheduled	% Canceled
May-00	217260	50078	6697	18.73%	2.99%
Oct-00	222200	58797	4408	20.92%	1.95%
May-01	216617	45598	10572	17.39%	4.65%
May-03	185714	50342	6541	21.33%	3.40%
Oct-03	196214	47699	4705	19.56%	2.34%

In general, there are a consistent percentage of unscheduled flights (around 20%) and canceled flights (around 3%). These data provide no indication as to the nature of the unscheduled/canceled flights. In Table 3.2 through 3.4 the above results are broken into the weight classes provided by ETMS. The weight classes again are heavy, large, and small.

Table 3.2. Results of ETMS/OAG Comparison (Heavy)

Date	Flown and Scheduled	Unscheduled	Canceled	% Unscheduled	% Canceled
May-00	23318	7046	1212	23.21%	4.94%
Oct-00	23565	7367	929	23.82%	3.79%
May-01	23799	5306	1393	18.23%	5.53%
May-03	18100	7549	1137	29.43%	5.91%
Oct-03	18885	7376	827	28.09%	4.20%

Table 3.3. Results of ETMS/OAG Comparison (Large)

Date	Flown and Scheduled	Unscheduled	Canceled	% Unscheduled	% Canceled
May-00	158881	15590	2754	8.94%	1.70%
Oct-00	163392	18243	1640	10.04%	0.99%
May-01	162628	12375	7316	7.07%	4.30%
May-03	147025	13803	2434	8.58%	1.63%
Oct-03	157106	13265	1347	7.79%	0.85%

Table 3.4. Results of ETMS/OAG Comparison (Small)

Date	Flown and Scheduled	Unscheduled	Canceled	% Unscheduled	% Canceled
May-00	35061	27442	2731	43.91%	7.23%
Oct-00	35243	33187	1839	48.50%	4.96%
May-01	30190	27917	1863	48.04%	5.81%
May-03	20589	28990	2970	58.47%	12.61%
Oct-03	20223	27058	2531	57.23%	11.12%

As can be seen, nearly half of all flights in the small category are unscheduled. In addition, the number of unscheduled flights in the small category is larger than the unscheduled flights in the combined heavy and large categories. However, since the results of this paper will be used in assessing global emissions to support SAGE, an analysis was performed to quantify the fuel burn contribution of each of the three weight categories as compared with the aggregate global fuel burn (Table 3.5).

Table 3.5. Aircraft Fuel Burn by Weight Class

Date	Fuel Burn (kg)			% Of Total Fuel Burn		
	Heavy	Large	Small	Heavy	Large	Small
May-00	288,354,873.51	601,681,860.50	9,573,914.64	32.05%	66.88%	1.06%
Oct-00	290,665,854.23	607,290,231.82	8,973,465.97	32.05%	66.96%	0.99%
May-01	248,240,868.89	606,442,669.63	9,399,602.25	28.73%	70.18%	1.09%
May-03	205,943,376.38	521,767,821.57	8,644,624.88	27.97%	70.86%	1.17%
Oct-03	213,687,209.25	551,247,967.97	9,187,861.19	27.60%	71.21%	1.19%

As can be seen, the Large and Heavy weight classes represent most of the global fuel burn (about 99%). Studies have shown that the uncertainty in SAGE fuel burn is approximately ±5% [FAA, 2005c][Lee, 2005], so fuel burn from small aircraft is well within SAGE uncertainty, at least in terms of fuel burn. Therefore, because of the general unpredictability observed in Table 3.4 for this weight class, small aircraft were removed for all further analyses presented herein.

4 Quantification of Unscheduled Flights

This section discusses the development of a methodology to account for unscheduled flights.

4.1 Correlation of Scheduled/Unscheduled Flights

First, the correlation between the number of scheduled and unscheduled departures by airport was determined for the five weeks of data (Table 4.1).

Table 4.1. Scheduled/Unscheduled Departures Correlation

Date	Correlation (Sched vs. Unsched)
May 00	0.801
October 00	0.843
May 01	0.770
May 03	0.649
October 03	0.674
Overall	**0.745**

Further analysis showed that the by-airport departure data is not equally distributed. Figures 4.1 through 4.5 show a distribution of airport sizes, as judged by the number of departures.

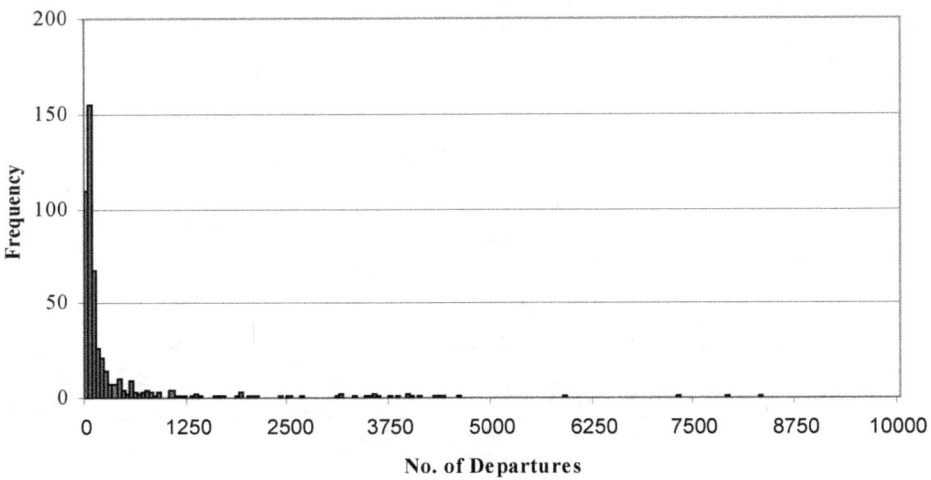

Figure 4.1. May 2000 Distribution of Airports by Size

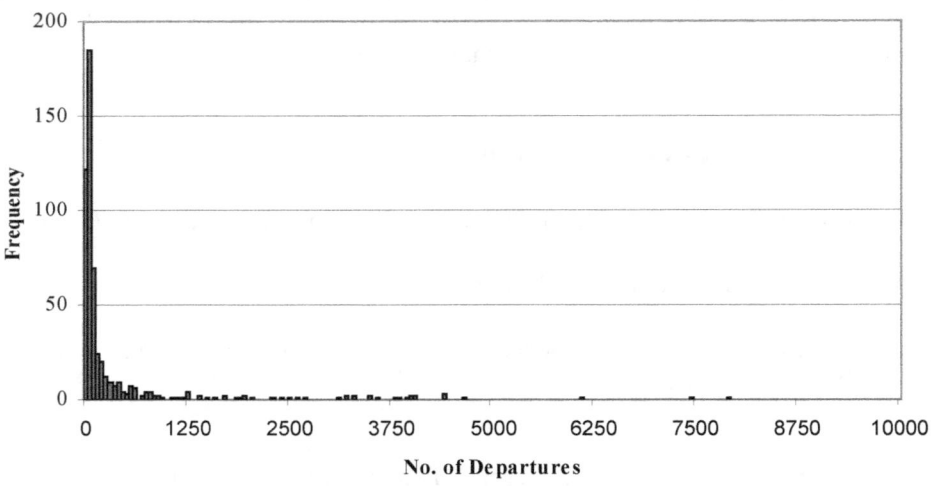

Figure 4.2. October 2000 Distribution of Airports by Size

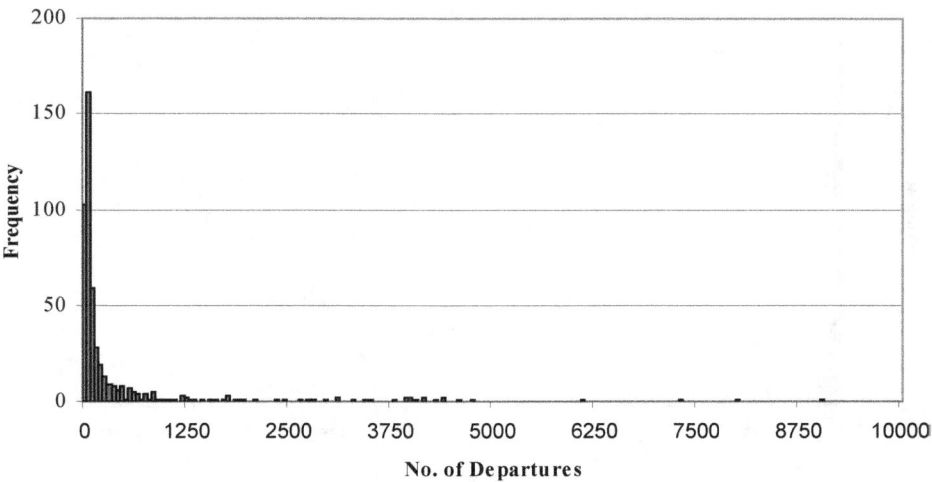

Figure 4.3. May 2001 Distribution of Airports by Size

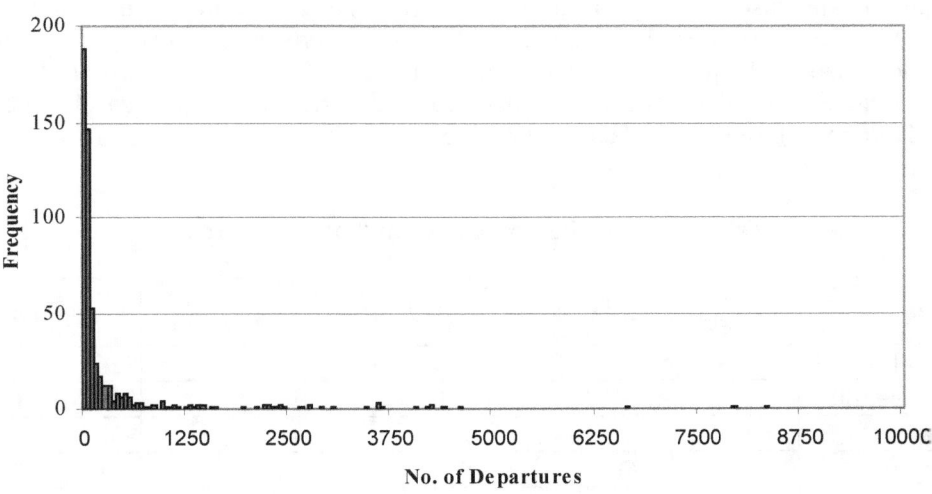

Figure 4.4. May 2003 Distribution of Airports by Size

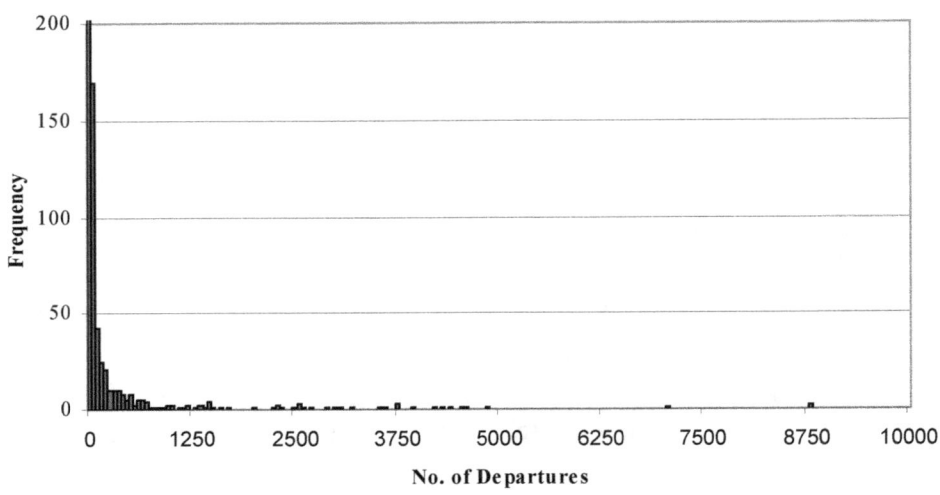

Figure 4.5. October 2003 Distribution of Airports by Size

To better understand the distribution graphics, all observed airports were broken into quartile ranges based on the number of departures. For example, 25% of all airports have more departures than the minimum but less than the 1st quartile amount, 25% of all airports have more departures than the 1st quartile amount but less than the 2nd quartile amount, etc. Table 4.2 presents the quartile ranges (and maximum and minimum) for the five data sets analyzed.

Table 4.2. Quartile Breakdown of Observed Airports

Date	Minimum	1st Quartile	Median	3rd Quartile	Maximum
May 00	0	1	42	185	8266
October 00	0	1	33	164	8540
May 01	0	2	44	199	9027
May 03	0	0	25	143	8311
October 03	0	0	21	134	8794
Overall	**0**	**0**	**32**	**166**	**9027**

As can be seen, there is a loading of the data to the smaller airports, with 32 being the median value, meaning 50% of all airports having fewer than 32 scheduled flights per week. The correlation was observed for each quartile; results are presented in Table 4.3

Table 4.3 Scheduled/Unscheduled Departures Correlation by Quartile

Date	Correlation			
	1st Quartile (Min to 1st Quartile)	2nd Quartile (1st Quartile to Median)	3rd Quartile (Median to 3rd Quartile)	4th Quartile (3rd Quartile to Max)
May 00	-0.156	0.124	0.287	0.711

October 00	-0.144	0.150	0.353	0.775
May 01	-0.234	0.113	0.238	0.667
May 03	n/a	0.081	0.308	0.533
October 03	n/a	0.087	0.100	0.572
Overall	**n/a**	**0.172**	**0.209**	**0.646**

As can be seen, the correlation coefficient for ¾ of all US airports is extremely poor, indicating that the number of unscheduled flights from smaller airports is more or less random. However, the majority of global fuel burn can be attributed to the larger airports. In Table 4.4, the total global fuel burn is compared to the fuel burn from flights departing from airports with 200 or more departures per week. (Note: For simplicity, the final application of the model will be for airports with 200 or more departures per week, however the analysis is based on airports with 166 or more departures per week, which corresponds precisely with the fourth quartile.) Table 4.4 shows that departures from all airports with 200 or more weekly departures account for more than 90% of the global fuel burn.

Table 4.4. Global Fuel Burn Contributions (Total vs. Large Airports)

Date	Total Fuel Burn (kg)	Fuel Burn (200 or more departures) (kg)	Percent Fuel Burn from 200 or more departures
May-00	3,399,233,395	3,077,241,353	90.53%
Oct-00	3,477,950,036	3,146,773,484	90.48%
May-01	3,170,301,920	2,863,565,972	90.32%
May-03	3,013,562,967	2,711,473,652	89.98%
Oct-03	3,178,252,001	2,887,919,706	90.87%

Consequently, the analysis presented herein will be limited to airports in the 4th Quartile (airports with 166 or more scheduled departures per week).

4.2 Scheduled vs. Unscheduled Flights

Presented below is a series of plots (Figures 4.6 through 4.10) of scheduled versus unscheduled flights for each of the five weeks included in the analysis; the x-axis represents the number of scheduled departures from an airport, and the y-axis represents the number of unscheduled departures from an airport.

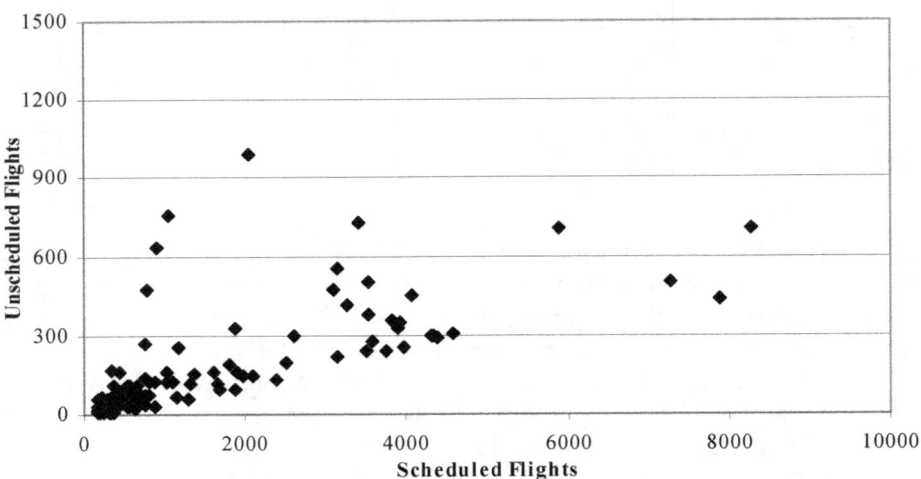

Figure 4.6. Graph of May 2000 Scheduled vs. Unscheduled

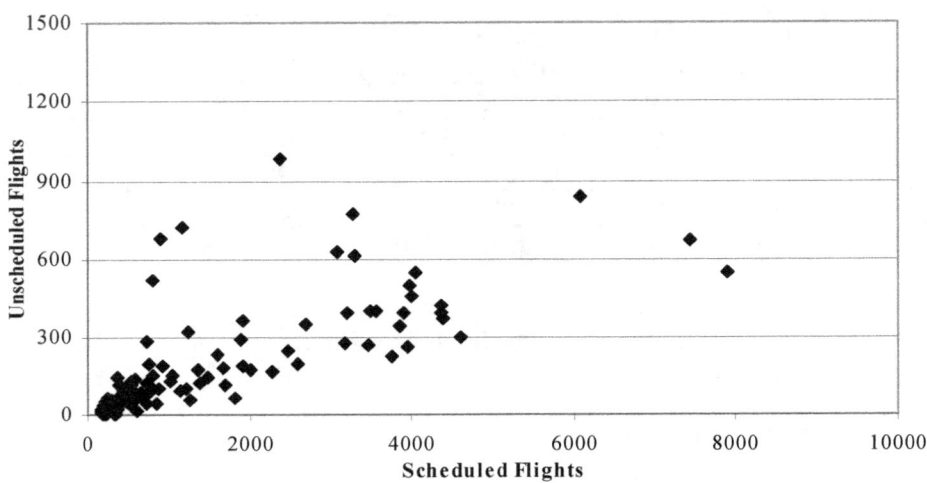

Figure 4.7. Graph of October 2000 Scheduled vs. Unscheduled

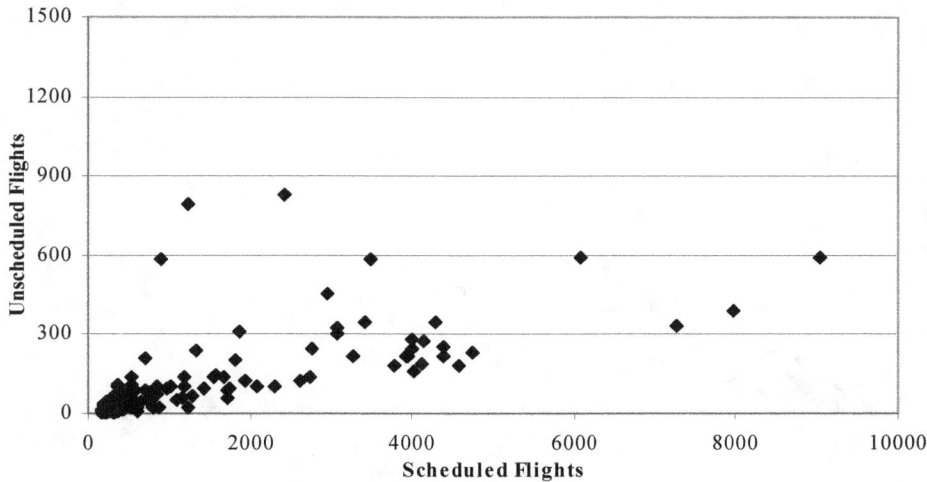

Figure 4.8. Graph of May 2001 Scheduled vs. Unscheduled

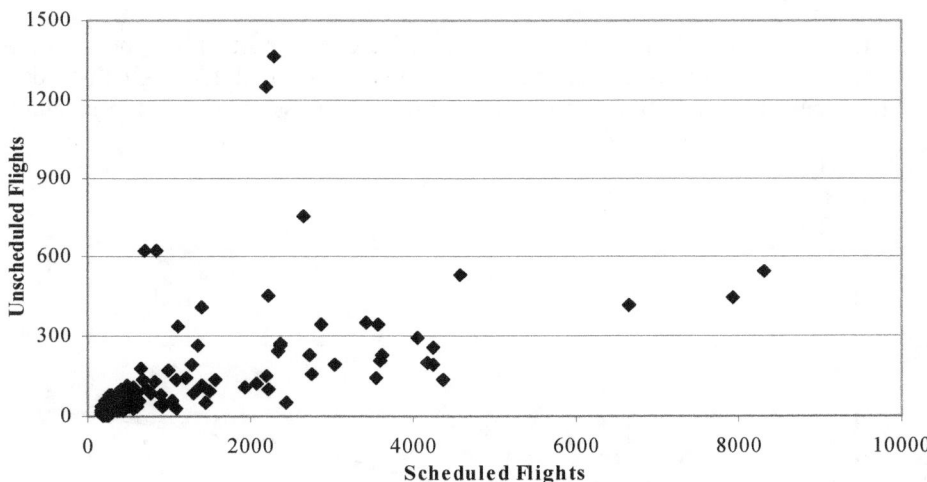

Figure 4.9. Graph of May 2003 Scheduled vs. Unscheduled

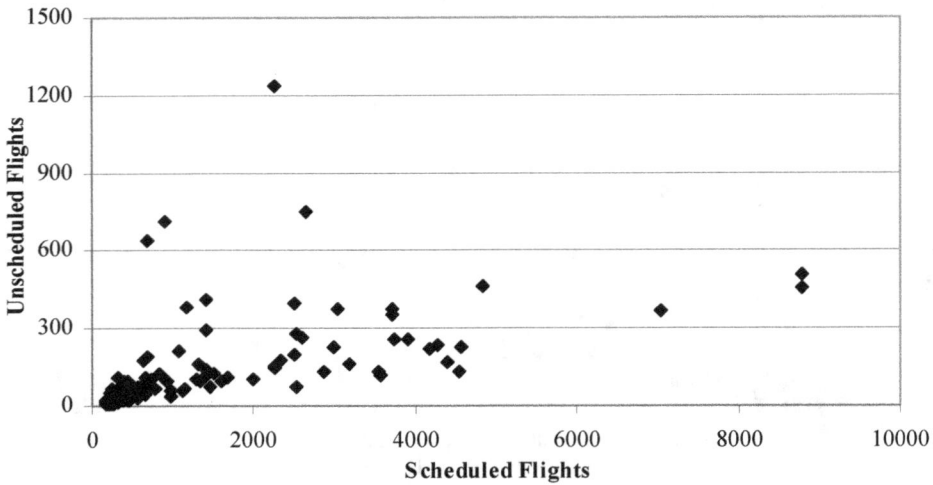

Figure 4.10. Graph of October 2003 Scheduled vs. Unscheduled

4.3 Final Relationship

With outlier airports removed (see Supplement A) and restricted to airports in the top ¼ quartile for the number of departures, a final relationship between scheduled and unscheduled flights was developed. All data from across the weeks were combined, and a second order regression was fit to the data (Figure 4.11).

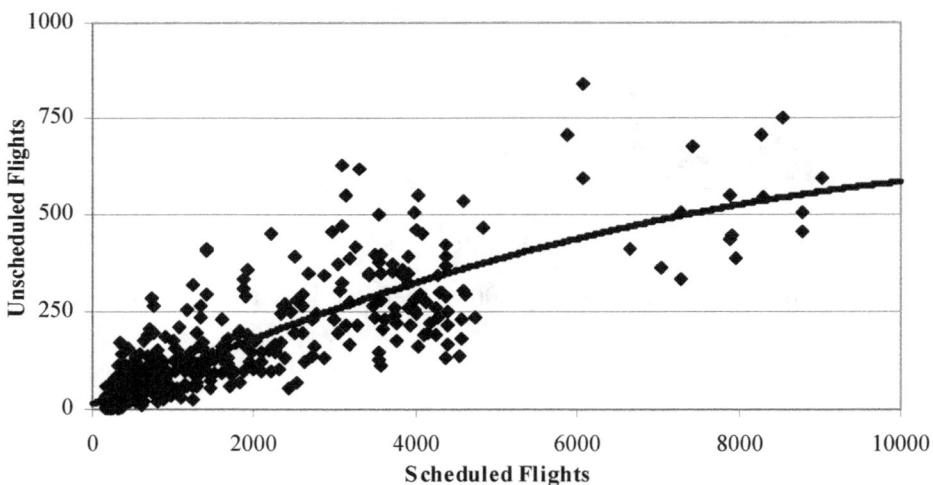

Figure 4.11. Overall Scheduled vs. Unscheduled 2nd Order Regression Fit

Regression: r-squared= 0.738
(Y = # of unscheduled flights, X = # of scheduled flights):
$$Y = 12.43653 + 0.09164\,X - 0.00000345 * X^2$$

Consequently, given the number of scheduled flights departing from a particular airport as provided by a source such as OAG, the number of unscheduled flights can be computed.

5 Quantification of Canceled Flights

This section discusses the development of a methodology to account for canceled flights.

5.1 Correlation of Scheduled/Canceled Flights

A similar analysis to that conduced for unscheduled flights was performed for canceled flights. As before, the correlation was broken down by quartile, based on departures (Table 5.1).

Table 5.1. Scheduled/Canceled Departures Correlation By Quartile

Date	Correlation			
	1st Quartile	2nd Quartile	3rd Quartile	4th Quartile
May 00	0.270	0.190	0.197	0.725
October 00	n/a	0.212	0.083	0.775
May 01	n/a	0.163	0.498	0.628
May 03	n/a	0.107	0.088	0.776
October 03	n/a	0.055	0.053	0.774
Overall	**n/a**	**0.157**	**0.089**	**0.699**

As before, the correlation is weak for ¾ of the data. Therefore, again, only the top ¼ busiest airports (by departures) were retained for the analysis.

5.2 Scheduled vs. Canceled Flights

Presented below is a series of plots (Figures 5.1 through 5.5) of scheduled versus canceled flights for each of the five weeks included in the analysis; the x-axis represents the number of scheduled departures from an airport, and the y-axis represents the number of canceled departures from an airport.

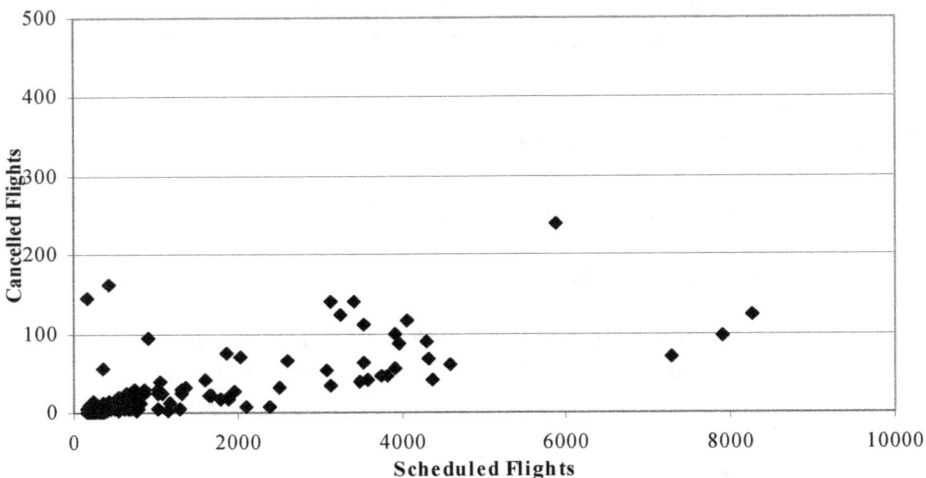

Figure 5.1. May 2000 Scheduled vs. Canceled

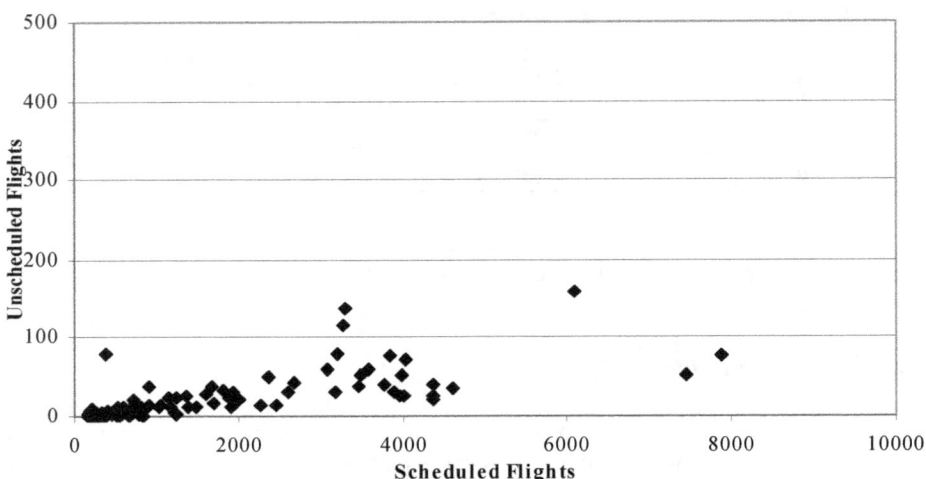

Figure 5.2. October 2000 Scheduled vs. Canceled

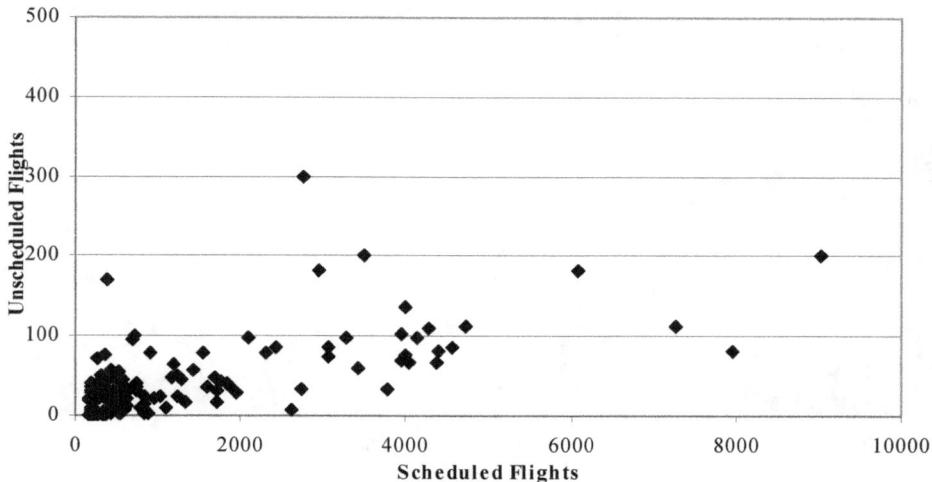

Figure 5.3. May 2001 Scheduled vs. Canceled

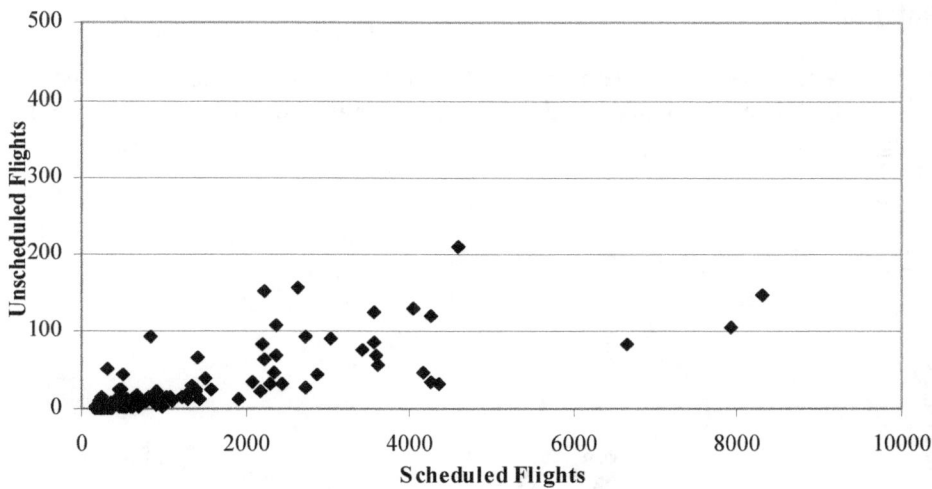

Figure 5.4. May 2003 Scheduled vs. Canceled

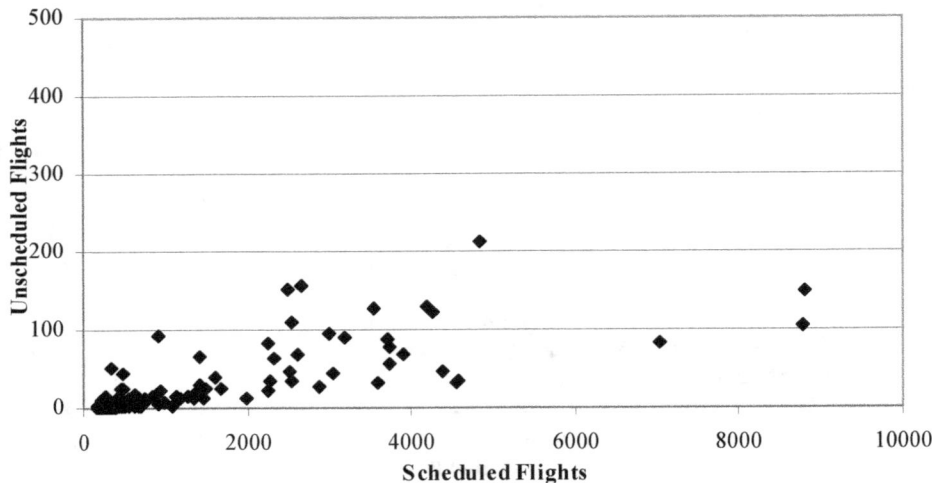

Figure 5.5. October 2003 Scheduled vs. Canceled

5.3 Final Relationship

With outlier airports removed (see Supplement A), a final relationship between scheduled and canceled flights was developed. All data from across the weeks were combined, and a second order regression was fit to the data (Figure 5.6).

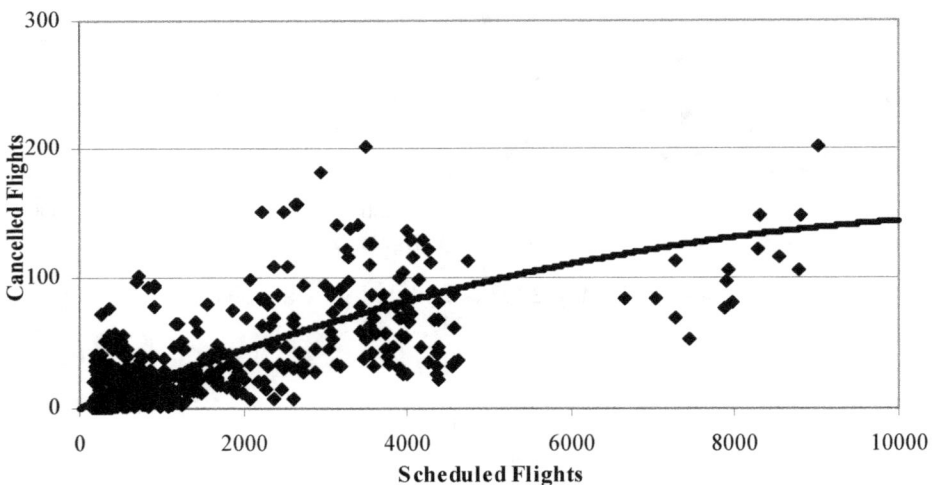

Figure 5.6. Overall Scheduled vs. Unscheduled 2nd Order Regression Fit

Regression: r-squared= 0.567
(Y = # of canceled flights, X = # of scheduled flights):
$$Y = 0.1728847 + 0.024352\,X + 0.000001 X^2$$

Consequently, given the number of scheduled flights departing from a particular airport as provided by a source such as OAG, the number of canceled flights can be computed.

6 Methodological Verification and Validation

This section quantifies the uncertainty associated with the developed methodology. The uncertainty is conducted using several independent data sets.

6.1 Verification of Combined Relationship

Table 6.1 presents an aggregate comparison between actual flights (from ETMS) and the scheduled flights (before applying the methodology presented herein) with the "calculated total flights" after applying the methodology presented herein. As can be seen, the difference before application ranges between 4 and 9 percent. After the methodology presented herein is applied the difference is between 0 and 3 percent.

Table 6.1. Overall Unscheduled/Canceled Adjustment

Date	Scheduled Flights	Actual Flights	% Diff (Sched vs Actual)	Calculated Total Flights	% Diff (Calculated vs Actual)
May-00	159316	171848	7.29%	170305	0.90%
Oct-00	154006	169287	9.03%	164774	2.67%
May-01	165260	171824	3.82%	176654	-2.81%
May-03	143496	153069	6.25%	153545	-0.31%
Oct-03	155735	165517	5.91%	166486	-0.59%

Next, in Figure 6.1, the results of data adjusted using the relationship is compared on an airport-by-airport basis. The x-axis represents the number of actual departures from a particular airport, and the y-axis represents the difference between the actual number of flights and either the number of scheduled flights or the number of calculated flights, as shown in the legend. A linear trend line and associated 95% confidence interval is also shown. Supplement B presents most of this data in tabular form. As can be seen, the calculated flights (after application of the adjustment methodology) tend to be randomly scattered around the zero difference line.

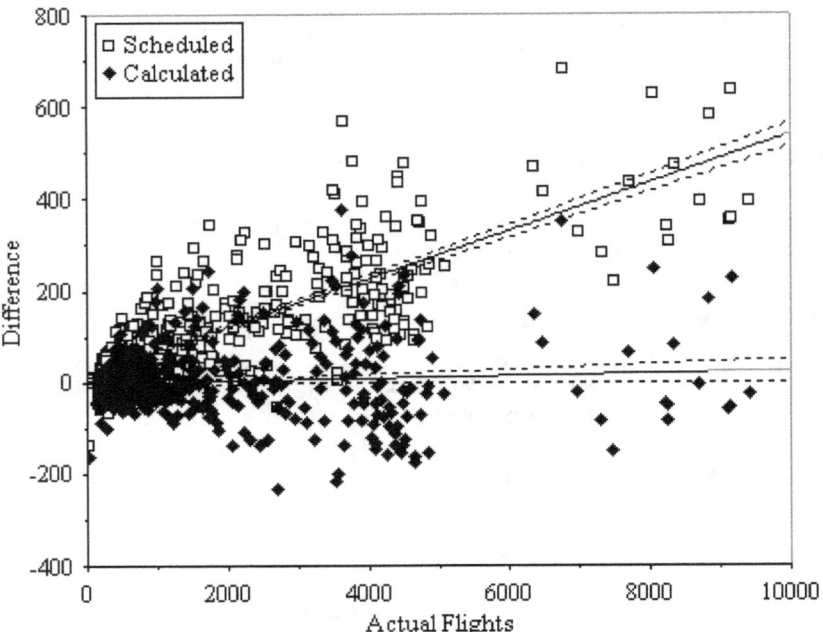

Figure 6.1. Scheduled/Canceled Flights vs. Actual Flights

Figure 6.2 in an expanded version of Figure 6.1, showing only airports with more than 166 but less than 2000 operations per week, which represents about 38% of operations taken into account, and about 31% of fuel burn.

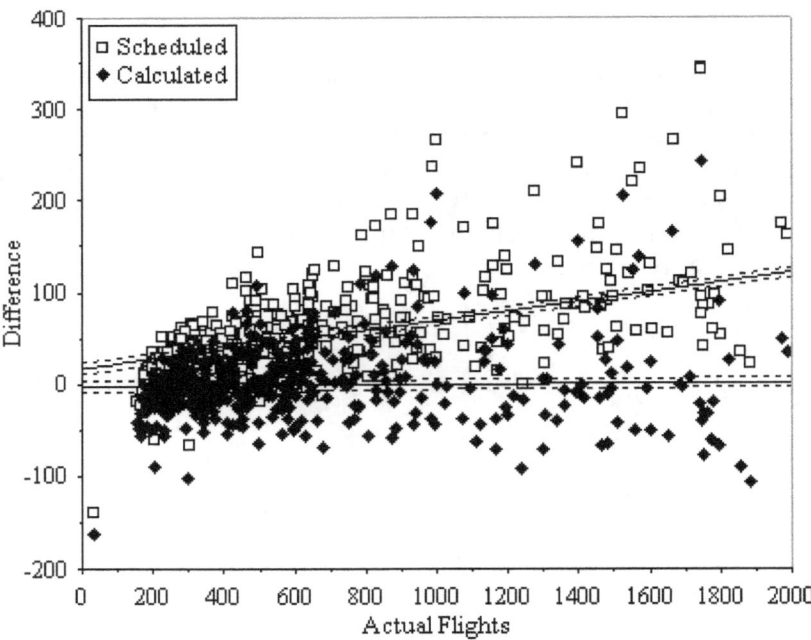

Figure 6.2. Scheduled/Canceled Flights (2000 or less) vs. Actual Flights

As can be seen, the difference between the actual and calculated number of flights is randomly clustered around zero.

6.2 Validation at Non-US Airports

To assess the validity of the methodology for application outside the US, comparisons were made with independent flight plan data not used to develop the methodology.

The first such data set used is a subset of ETMS not used in the original model development. Since ETMS covers all of North America, and the methodology was developed using only data from US airports, data from Canadian airports could be used to conduct an independent assessment of the method. Table 6.2 shows the results of the comparison, focusing on the three busiest airports in Canada (Toronto: CYYZ, Vancouver: CYVR, and Montreal: CYUL):

Table 6.2. Canadian Airport Analysis

Date	Airport Code	Sched Flights	Actual Flights	% Diff	Calc. Flights	% Diff
May-00	CYYZ	3334	3646	8.56%	3543	2.81%
May-00	CYVR	1846	2106	12.35%	1974	6.26%
May-00	CYUL	1370	1517	9.69%	1470	3.11%
Oct-00	CYYZ	3543	3884	8.78%	3763	3.12%
Oct-00	CYVR	1920	2137	10.15%	2052	3.96%
Oct-00	CYUL	1433	1609	10.94%	1537	4.50%
May-01	CYYZ	3428	3645	5.95%	3642	0.08%
May-01	CYVR	1926	2075	7.18%	2059	0.78%
May-01	CYUL	1309	1397	6.30%	1405	-0.58%
May-03	CYYZ	2629	2852	7.82%	2801	1.78%
May-03	CYVR	1472	1579	6.78%	1578	0.06%
May-03	CYUL	1205	1328	9.26%	1295	2.50%
Oct-03	CYYZ	2965	3196	7.23%	3155	1.28%
Oct-03	CYVR	1506	1611	6.52%	1614	-0.19%
Oct-03	CYUL	1373	1504	8.71%	1473	2.06%
Total		31259	34086	8.29%	33362	2.12%

As can be seen the model results in a more precise accounting of flights for all weeks and airports analyzed. The overall total % difference is reduced from 8.29% to 2.12%.

As part of the SAGE development effort, the FAA and the Volpe Center have been working with the Eurocontrol Experimental Center to better characterize global fleet movements [Michot et. al, 2004]. As part of this effort both U.S. and European data have been exchanged by the organizations. For the next comparison, the results of the methodology as applied to the OAG are compared with the Eurcontrol data for a selected week in October 2003, and for the top 20 busiest airports in Europe. The comparisons are presented in Table 6.3.

Table 6.3. European Airport Analysis

Airport Name	Airport Code	Sched Flights	Actual Flights	% Diff	Calc. Flights	% Diff
Heathrow, London, UK	EGLL	4431	4464	0.74%	4693	-5.14%
Charles-De-Galle, Paris, France	LFPG	4429	4887	9.37%	4691	4.01%
Frankfurt, Germany	EDDF	3876	4098	5.42%	4112	-0.35%
Amsterdam, Netherlands	EHAM	2996	3372	11.15%	3188	5.46%
Madrid, Spain	LEMD	3545	3637	2.53%	3765	-3.52%
Gatwick, London	EGKK	1665	2280	26.97%	1783	21.82%
Rome, Italy	LIRF	2804	2919	3.94%	2986	-2.28%
Orly, Paris, France	LFPO	1939	2089	7.18%	2073	0.79%
Munich, Germany	EDDM	3022	3145	3.91%	3215	-2.23%
Zurich, Switzerland	LSZH	1719	2037	15.61%	1840	9.69%
Brussels, Belgium	EBBR	1348	1733	22.22%	1447	16.53%
Milan, Italy	LIMC	1757	1936	9.25%	1880	2.90%
Barcelona, Spain	LEBL	2386	2519	5.28%	2545	-1.03%
Palma De Mallorca, Spain	LEPA	1169	1678	30.33%	1257	25.11%
Manchester, UK	EGCC	1355	1898	28.61%	1454	23.40%
Stockholm, Sweden	ESSA	1878	2046	8.21%	2008	1.86%
Copenhagen, Denmark	EKCH	2220	2383	6.84%	2370	0.56%
Dusseldorf, Germany	EDDL	1457	1593	8.54%	1562	1.94%
Oslo, Norway	ENGM	1635	1718	4.83%	1751	-1.91%
Vienna, Austria	LOWW	1615	1763	8.39%	1730	1.90%
Total		47246	52195	9.48%	50350	3.53%

As was the case for Canada, on an aggregate basis, the methodology results in a substantial improvement. The overall percentage difference is reduced from 9.48% to 3.53%; and for 18 of the 20 airports, the methodology results in a more accurate representation of the number of flights.

As a last point of comparison, actual flight data were obtained for a two-week period (October 6th-October 19th 2003) for an airport in South America: Dep. L. E. Magalhaes International, Salvador, Brazil (Airport Code: SBSV). The results of the comparison are shown in Table 6.4.

Table 6.4. Brazilian Airport Results

Date	Scheduled Flights	Actual Flights	% Diff	Calculated Flights	% Diff
Oct. 6th-12th 2003	453	507	10.65%	495	2.37%
Oct. 13th-19th 2003	453	498	9.04%	495	0.60%

As can be seen, a substantial improvement is realized when the methodology is applied.

8 Conclusions

In any SAGE analysis requiring flight data where ETMS or an equivalent source of radar-quality data is not available, and OAG is relied upon, it is recommended that the methodology presented herein be utilized to adjust the OAG data. The specific final equation that should be applied is as follows:

(predicted flights = scheduled flights + predicted unscheduled − predicted canceled)
(Y = # of predicted flights, X = # of scheduled flights):

$$Y = 12.2636453 + 1.067288 X + 0.00000245 X^2$$

A comprehensive validation of these equations was conducted for all airports with more than 166 weekly operations. For simplicity, it is recommended that they be applied for to airports with more than 200 weekly operations. As Table 4.4 demonstrated, airports with more than 200 weekly operations account for about 90% of global fuel burn from commercial aviation. The relationship developed will result in a substantial improvement in the number of unscheduled/canceled flights when applied to OAG scheduled data.

References

CNN. "Comair Pilots OK Contracts; Flights Resume July 2." CNN. http://archives.cnn.com/2001/TRAVEL/NEWS/06/22/comair.strike/index.html. June, 2001.

Duchin, Winston. "More Radar Problems for Airport." MSNBC. http://archives.californiaaviation.org/airport/msg07892.html. May, 2000.

Federal Aviation Administration (FAA). "Official Airline Guide." FAA Office of Aviation Policy and Plans (APO). http://apo.faa.gov/apo_home.asp. 2005.

Federal Aviation Administration (FAA). "System for assessing Aviation's Global Emissions (SAGE), Version 1.5, Technical Manual." FAA-EE-2005-01. August 2005.

Federal Aviation Administration (FAA). "System for assessing Aviation's Global Emissions (SAGE), Version 1.5, Validation Assessment and Model Assumptions." FAA-EE-2005-03. August 2005.

Lee, J.J., "Modeling Aviation's Global Emissions, Uncertainty Analysis, and Applications to Policy," PhD Thesis, Massachusetts Institute of Technology, February 2005.

Michot, Sophie, Ted Elliff, Gregg G. Fleming, Brian Kim, Curtis A. Holsclaw, Maryalice Locke, and Angel Morales. "Flight Movement Inventory: SAGE-AERO2K." Air Traffic Control Quarterly. Vol 12(2) 125-145. 2004.

Volpe National Transportation Systems Center/US DOT. "Enhanced Traffic Management System (ETMS), Functional Description, Version 7.6." Report Number VNTSC-DTS56-TMS-002. August 2003.

Supplement A: Analysis of Outlying Airports

As discussed in Section 4.3 and 5.3, a few outlier airports were removed from the data set used to develop the final recommended methodology presented herein. This Supplement discusses the rationale for removing these airports.

A.1 Unscheduled Flights Outliers

First, there were several airports that were identified with an extremely high number of unscheduled flights. The outlying airports are identified for the two weeks analyzed in 2000 (Figures A.1 and A.2).

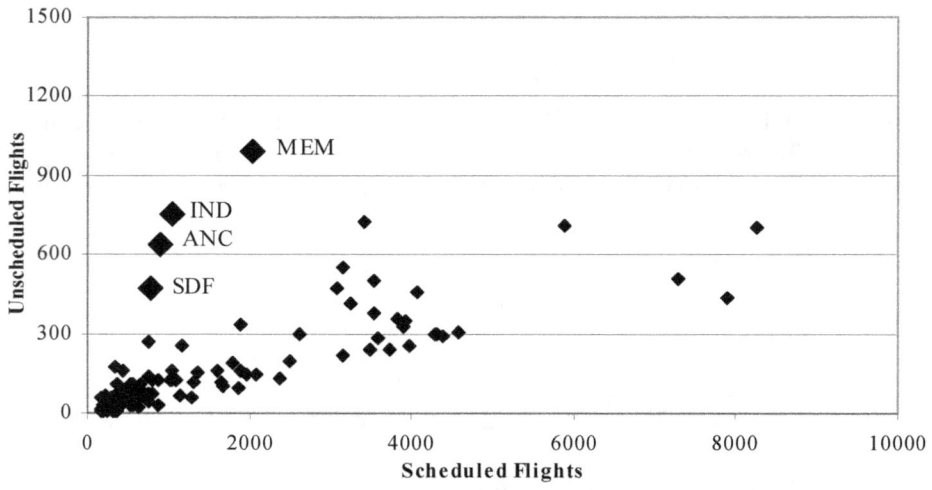

Figure A.1. Graph of May 2000 Scheduled vs. Unscheduled (Outliers)

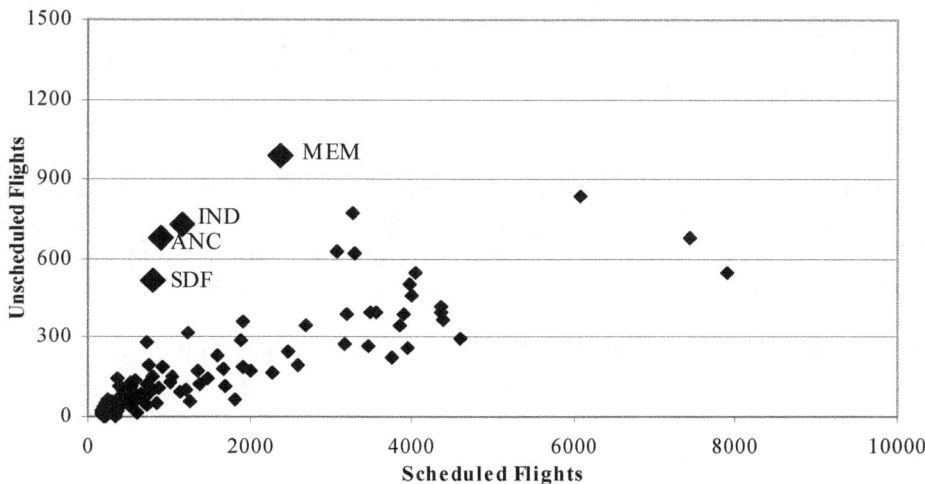

Figure A.2. Graph of October 2000 Scheduled vs. Unscheduled (Outliers)

These airports are: MEM (Memphis, Tennessee), IND (Indianapolis, Indiana), Louisville, Kentucky (SDF), and Anchorage, Alaska (ANC). MEM, IND, and ANC are hubs for either FedEx or UPS, which operate as the main sorting facilities for these large cargo carriers. The basic principle of the hubs is, across the country, the aircraft of FedEx and UPS are loaded and sent to the major hubs. Once there, the cargo is sorted, loaded back onto the aircraft, and shipped out. So, for example, a shipment from New York to Los Angeles is loaded onto an aircraft, flown to Memphis, and sorted onto an aircraft bound for Los Angeles. Memphis is the major hub of FedEx with a "mini-hub" in Indianapolis, and Louisville is the major hub of UPS. Anchorage acts as a "mini-hub" for FedEx, UPS, and a number of other small cargo carriers. The problem in predicting unscheduled flights for these airports is that a vast majority of cargo flights are unscheduled; they operate on a "fly as needed" basis. Tables A.1 and A.2 show the percentage of cargo flights at these four airports as compared to all other airports observed, (with more than 166 departures and located in the US).

Table A.1. Unscheduled Cargo Flights for May 2000

Airport Code	Unscheduled Flights	Unscheduled Cargo Flights	Percentage of Cargo Flights
MEM	992	848	85.48%
IND	754	477	63.27%
SDF	473	443	93.66%
ANC	636	295	46.38%
All Airports	**19679**	**7648**	**38.86%**

Table A.2. Unscheduled Cargo Flights for October 2000

Airport Code	Unscheduled Flights	Unscheduled Cargo Flights	Percentage of Cargo Flights

MEM	985	821	83.35%
IND	728	480	65.93%
SDF	522	476	91.19%
ANC	678	288	42.48%
All Airports	**21746**	**7329**	**33.70%**

As can be seen, cargo flights account for a substantially higher percentage of flights at these four airports, particularly at MEM, SDF and IND. Most airports have only a modest number of cargo operations, and thus have a minimal impact on the number of unscheduled flights. However, when an airport operates as a cargo hub, a majority of the flights will by definition be cargo.

Next, the outlying airports for the week analyzed in 2001 are presented in Figure A.3.

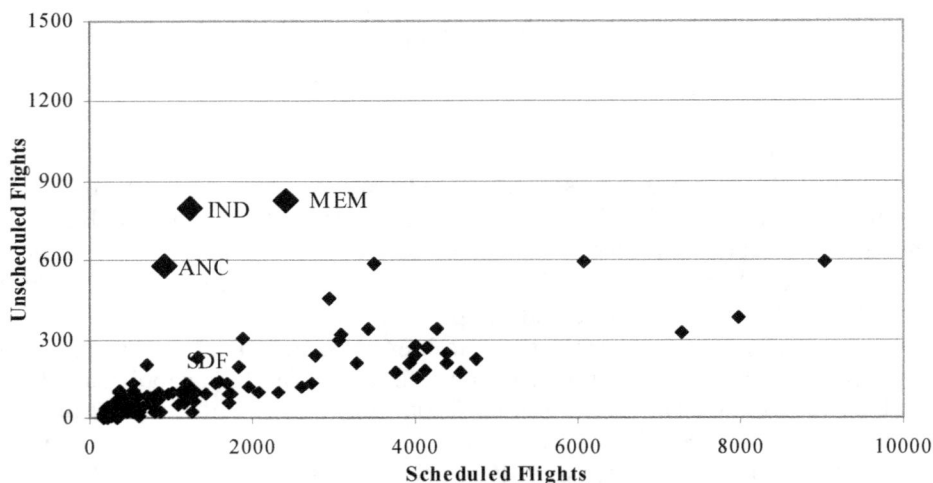

Figure A.3. Graph of May 2001 Scheduled vs. Unscheduled (Outliers)

As can be seen, MEM, IND and ANC were clear outliers again. SDF (Louisville) was not a clear outlier for this week.

Finally, the two analyzed weeks in the year 2003 are shown in Figures A.4 and A.5.

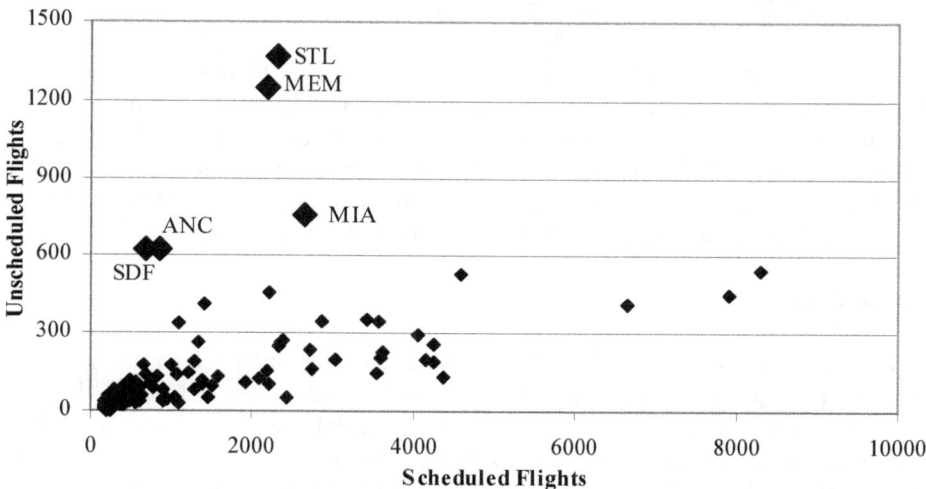

Figure A.4. Graph of May 2003 Scheduled vs. Unscheduled (Outliers)

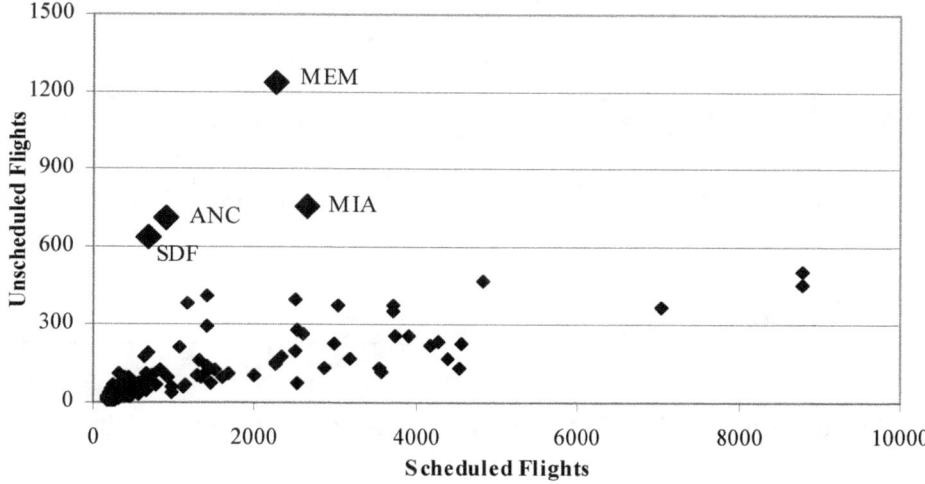

Figure A.5. Graph of October 2003 Scheduled vs. Unscheduled (Outliers)

Memphis, Louisville, and Anchorage are once again outlying airports. Also, MIA (Miami) is shown to be a new outlier; Miami does serve as a "mini-hub" for UPS and FedEx.

Finally, for the month of May 2003, STL (St. Louis) is somewhat of a unique outlier; it is not noticeably different in other months and years observed, but for this particular month it registers as having one of the highest number of unscheduled flights. Upon further analysis, a few sub-carriers of Delta airlines: Chautauqua and Trans States Airlines represent 1,075 of the 1,504 flights found to be operating unscheduled from St. Louis. This anomaly does not appear in any of the other analyzed data sets.

In conclusion, MEM, SDF, IND, MIA and ANC were omitted from ALL data sets, while STL (for the unique reasons cited above) was only omitted from the May 2003 data set.

A.2 Approximation for Airports Dominated by Cargo Flights

The general methodology presented herein is for airports with over 200 weekly flights, that do not include a substantial number of cargo operations. The previous section brought to light a number of airports with a large amount of unscheduled flights, most notably high-cargo volume airports. These airports have been excluded from the previous analysis presented herein. However, a methodology is required to adjust scheduled operations at these airports, as well as cargo-dominated airports worldwide. This section presents a recommended methodology for airports dominated by cargo flights.

A.2.1. U.S. Cargo Airports

To support this analysis a year's worth of ETMS and OAG data was analyzed for the five outlying airports, and the percentage of additional flights was found on a week-by-week basis. Figures A.6 through A.10 show the percentage of additional flights on a week by week basis when ETMS is compared to OAG for each of the outlier airports, MEM, IND, SDF, ANC and MIA, respectively. These graphics show the percentage of additional flights, meaning, it is a combination of the number of unscheduled and canceled flights.

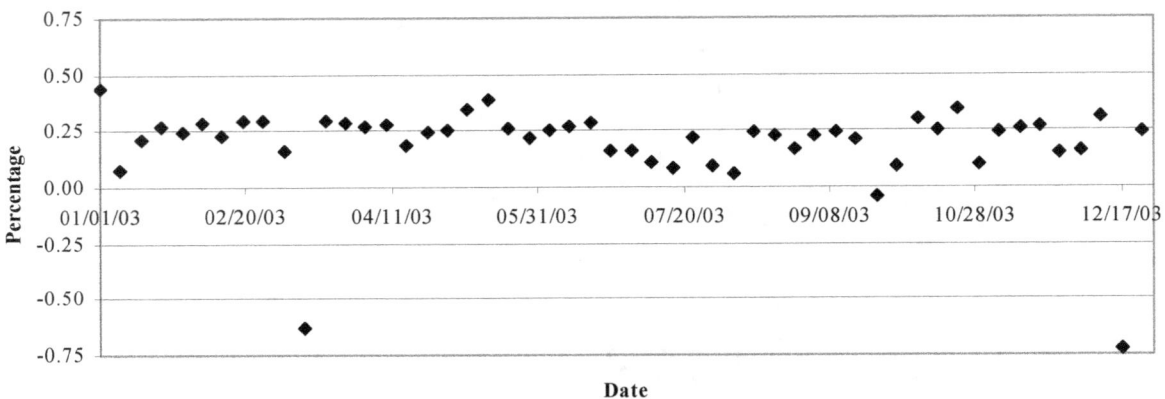

Figure A.6. 2003 MEM (Memphis) Percentage of Additional Flights

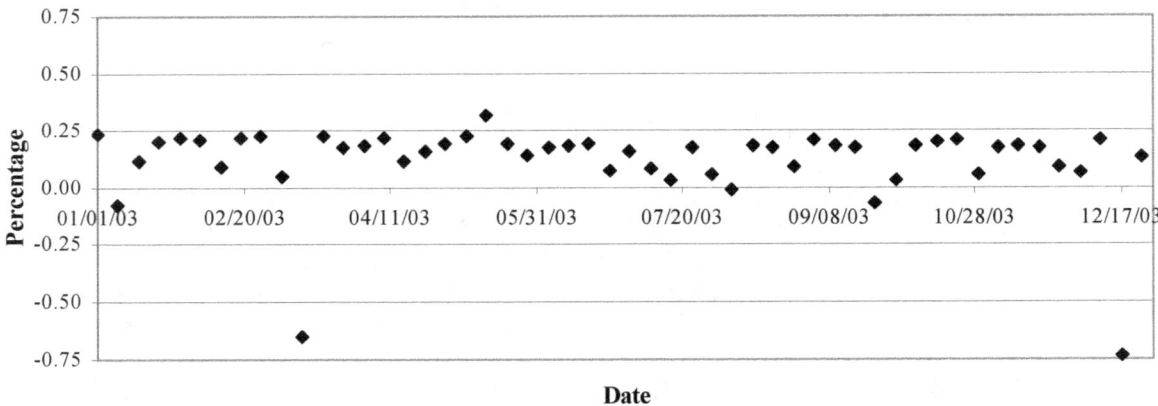

Figure A.7. 2003 IND (Indianapolis) Percentage of Additional Flights

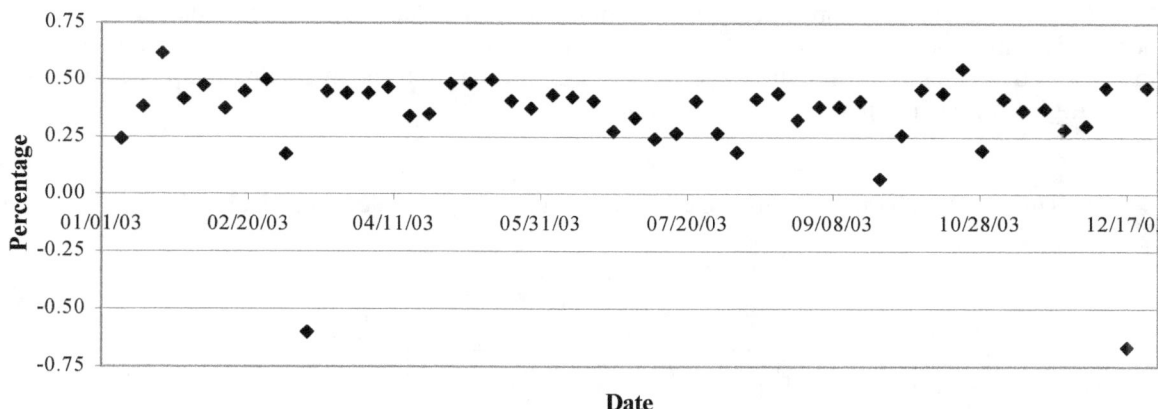

Figure A.8 2003 SDF (Louisville) Percentage of Unscheduled Flights

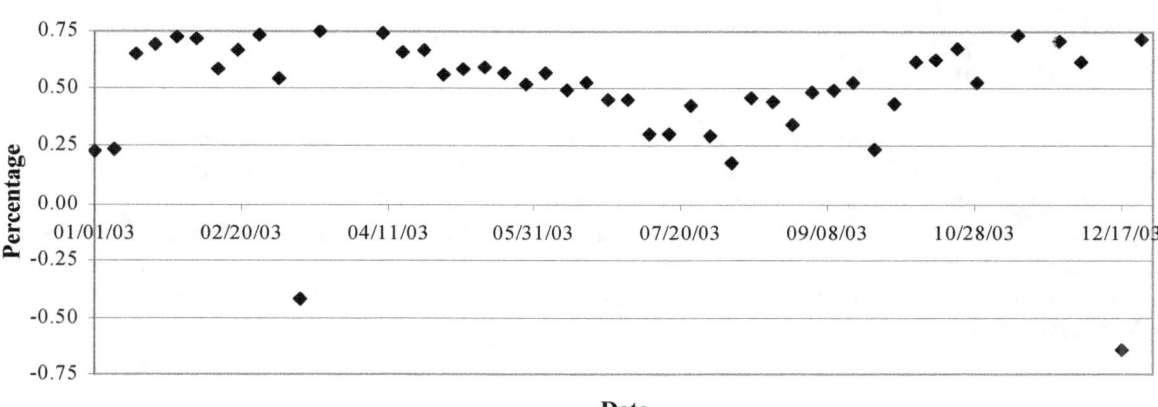

Figure A.9. 2003 ANC (Anchorage) Percentage of Unscheduled Flights

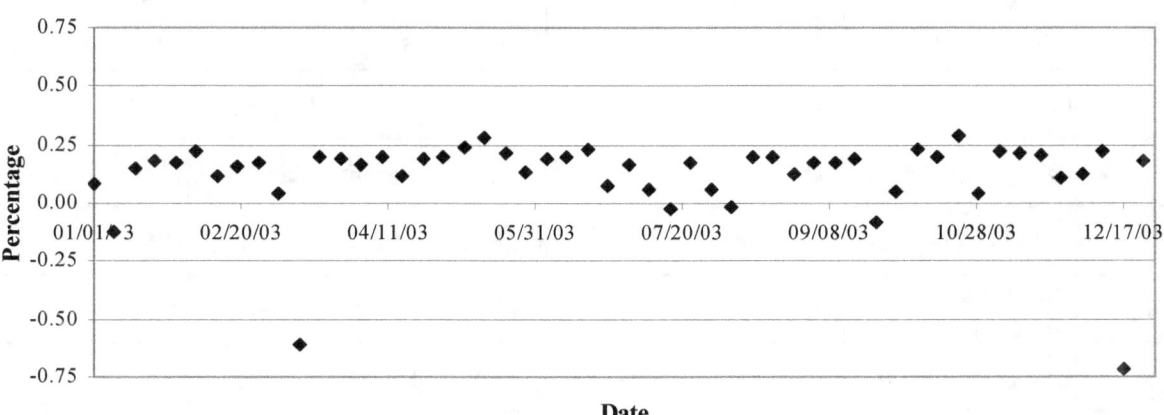

Figure A.10. 2003 MIA (Miami) Percentage of Unscheduled Flights

As can be seen, two specific weeks are consistent outliers throughout each data set: 3/12/2003-3/18/2003 and 12/17/2003-12/232003. Upon further investigation, the ETMS data set does not contain any flights for specific days in these weeks. The outliers are not a result of the airports themselves, but rather missing data in the data set. Removing these weeks from the analysis, and averaging the percentage of additional flights for each airport we arrive at Table A.3.

Table A.3. Average Percentage of Additional Flights for Selected Airports, 2003

Airport	Average % Additional Flights (Actual)
MEM	22.2%
IND	14.4%
SDF	38.5%
ANC	55.0%
MIA	14.7%

A.2.2. Non-U.S. Cargo Airports

Given the simple, airport-specific, factor-based approach presented in Section A.2.1 for the five U.S. airports, it was recognized that similar factors would be needed for non-U.S. Cargo Airports. The first step was to analyze the number of ETMS cargo flights that operate internationally. Using the 2003 ETMS data, the non-US airports with the highest number of ETMS cargo flights were found and the results are presented in Tables A.4 and A.5 (note: as ETMS covers North America, a larger number of North American flights appear. Therefore, two tables were made: Non-US North American Airports and Non-North American Airports).

Table A.4 Cargo Airports (Non-North American) 2003

Airport Code	Airport Location	No. of Yearly ETMS Cargo Flights (2003)
RJAA	Tokyo, Japan	1577
EGSS	London, UK	1261
LFPG	Paris, France	1157
EDDK	Cologne, Germany	1062
EDDL	Dusseldorf, Germany	642
EGNX	East Midlands, UK	620
RKSI	Seoul, Korea	619

Table A.5 Cargo Airports (North American) 2003

Airport Code	Airport Location	No. of Yearly ETMS Cargo Flights (2003)
TJSJ	San Juan, Puerto Rico	1690
MMUN	Cancun, Mexico	1317
CYYZ	Toronto, Canada	1310
CYMX	Montreal, Canada	1057
CYYC	Calgary, Canada	989
CYHM	Hamilton, Canada	805
CYVR	Vancouver, Canada	800

RJBB	Osaka, Japan	601	MDSD	Santo Domingo, Dominican Republic	535
VHHH	Hong Kong	474	MDPC	Higuey, Dominican Republic	513
SEQU	Quito, Ecuador	417	CYWG	Winnipeg, Canada	500
SBKP	Campinas, Brazil	335	MMPR	Puerto Vallarta, Mexico	487
EDDF	Frankfurt, Germany	323	MMMX	Mexico City, Mexico	302
EHAM	Amsterdam, Netherlands	304	MDPP	Puerto Plata, Dominican Republic	274
ZSPD	Shanghai, China	294	MHLM	San Pedro Sula, Honduras	256
YSSY	Syndey, Australia	243	MGGT	Guatemala City, Guatemala	250
EDDM	Munich, Germany	217	MSLP	San Salvador, El Salvador	240
SVVA	Valencia, Venezuela	217	MPTO	Panama City, Panama	114
EBBR	Brussels, Belgium	174	CYOW	Ottawa, Canada	91
LIMC	Milan, Italy	153	MMTO	Toluca, Mexico	73
EBOS	Ostend, Belgium	134	MUVR	Varadero, Cuba	68

A.2.2.1. North American Airports

First, the North American airports listed above were analyzed for the entire year of 2003. Any airport that did not have at least 166 weekly operations (of all types, not just cargo) was discarded since it would not be included in the methodology presented herein. The remaining airports were: TJSJ (San Juan), MMUN (Cancun), CYYZ (Toronto), CYYC (Calgary), CYVR (Vancouver), MDSD (Santo Domingo), CYWG (Winnipeg), MMMX (Mexico City), and CYOW (Ottawa). Figures 7.11 through 7.19 show the percentage of additional flights on a week by week basis when ETMS is compared to OAG for each of the remaining airports. As was the case for the U.S. cargo airport analysis, these graphics show the percentage of additional flights, meaning, it is a combination of the number of unscheduled and canceled flights.

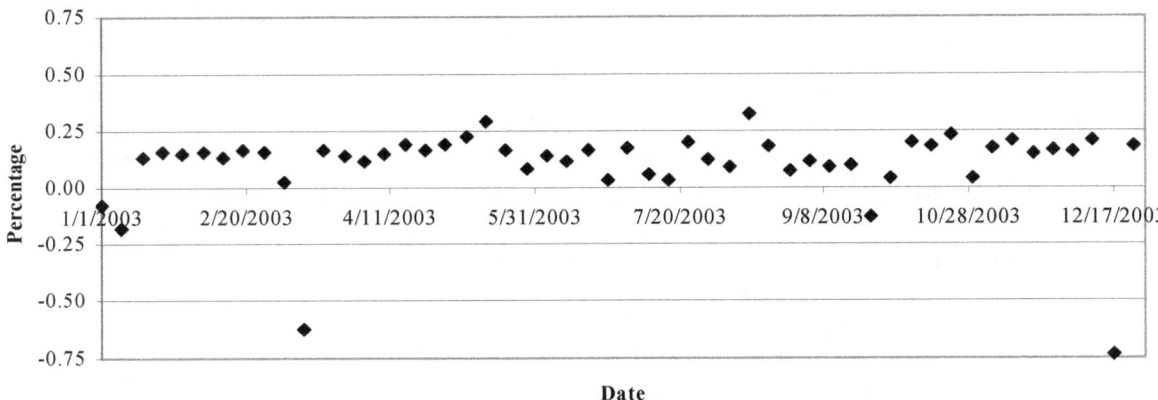

Figure A.11. 2003 TJSJ (San Juan) Percentage of Additional Flights

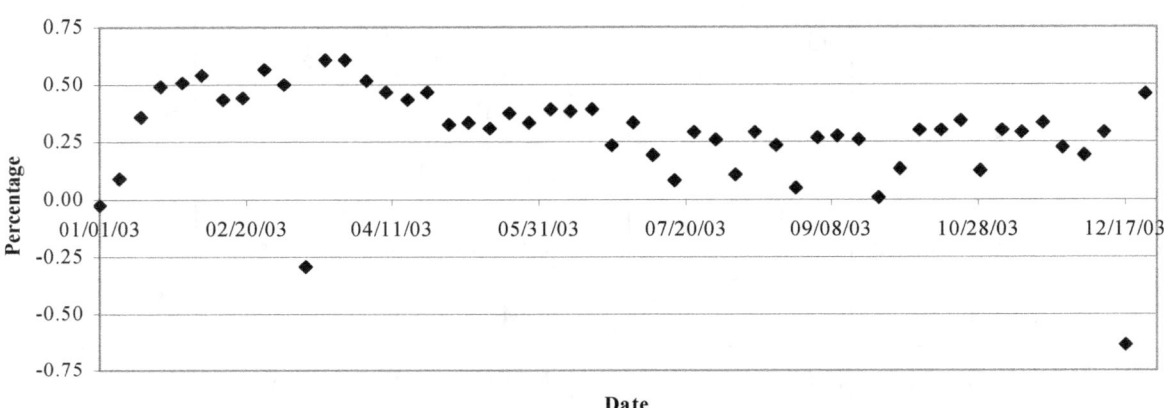

Figure A.12. 2003 MMUN (Cancun) Percentage of Additional Flights

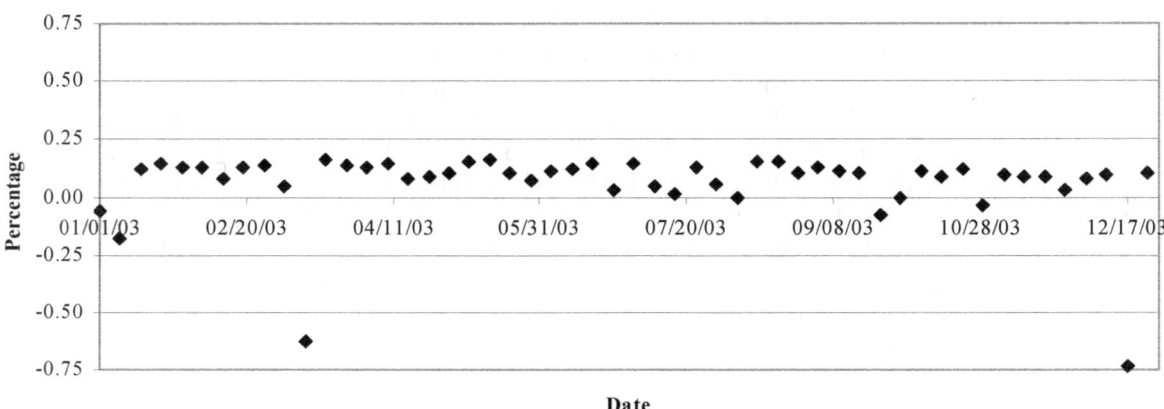

Figure A.13. 2003 CYYZ (Toronto) Percentage of Additional Flights

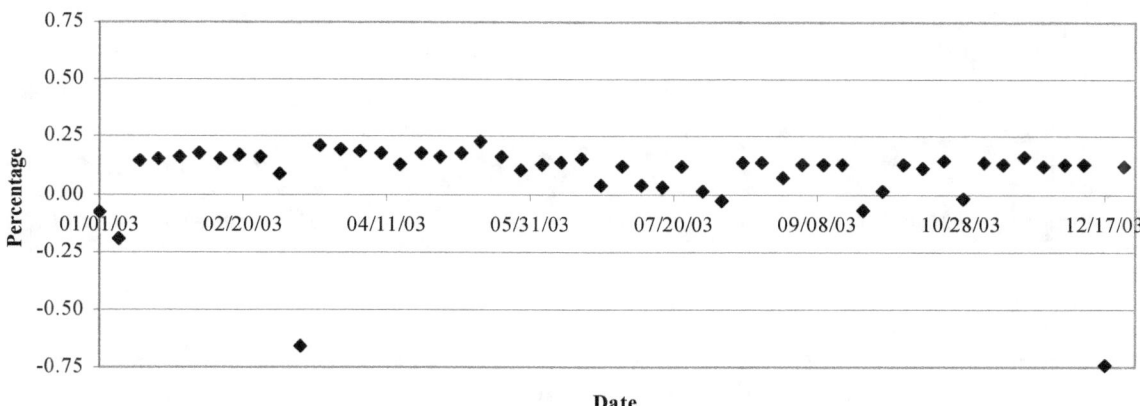

Figure A.14. 2003 CYYC (Calgary) Percentage of Additional Flights

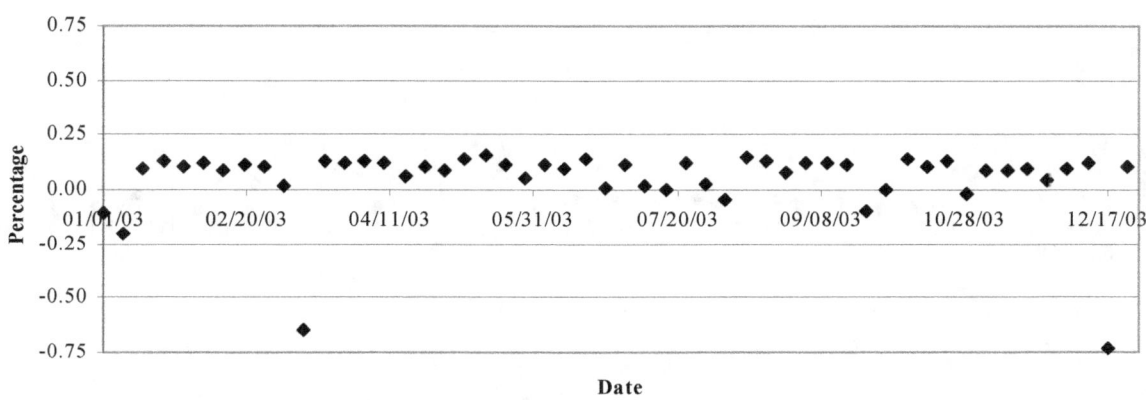

Figure A.15. 2003 CYVR (Vancouver) Percentage of Additional Flights

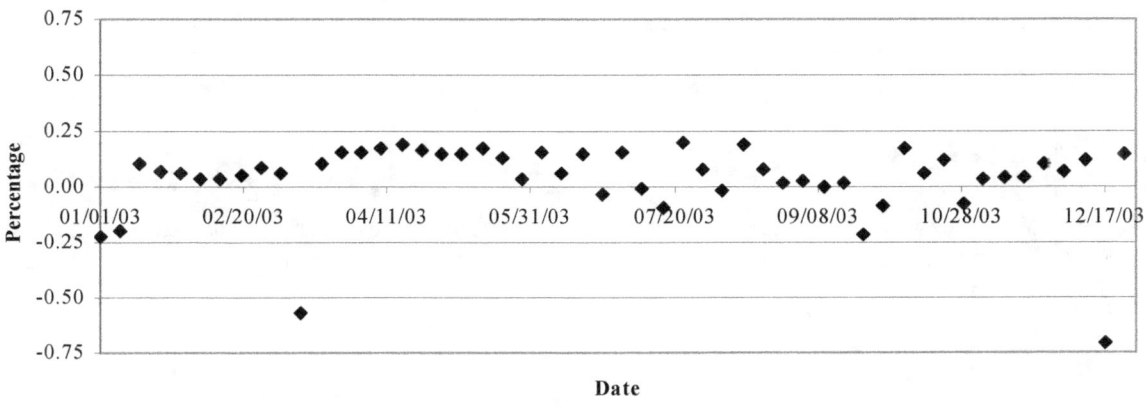

Figure A.16. 2003 MDSD (Santo Domingo) Percentage of Additional Flights

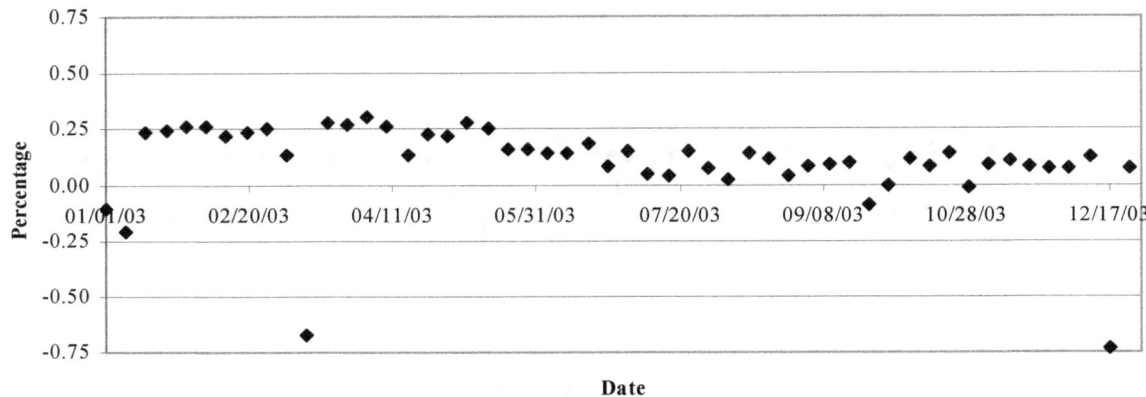

Figure A.17. 2003 CYWG (Winnipeg) Percentage of Additional Flights

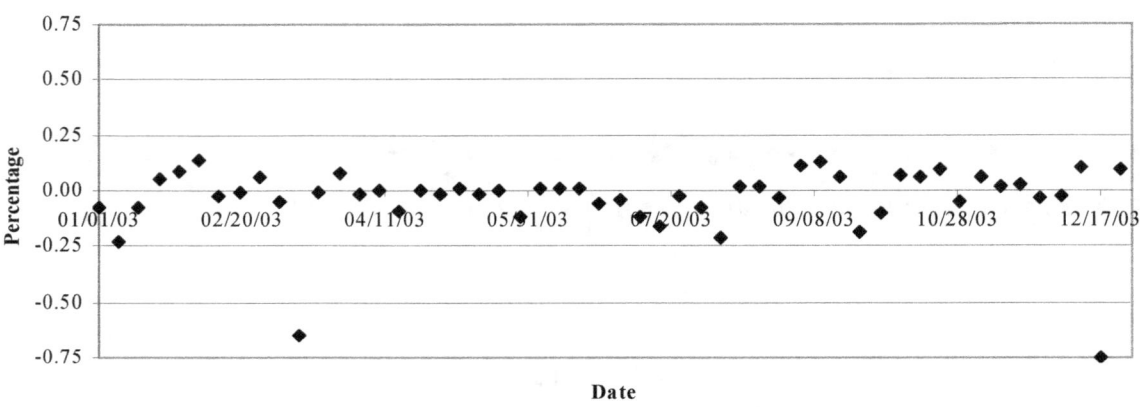

Figure A.18. 2003 MMMX (Mexico City) Percentage of Additional Flights

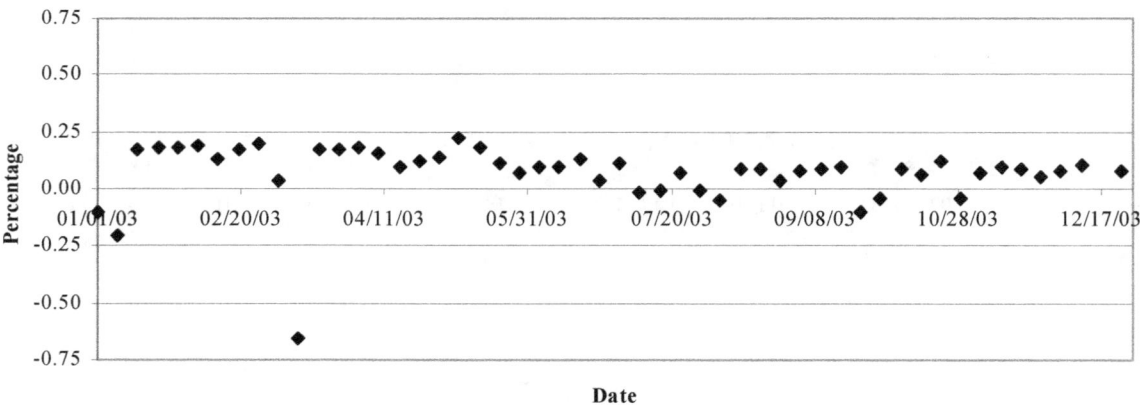

Figure A.19. 2003 CYOW (Ottawa) Percentage of Additional Flights

As was the case in the US cargo analysis, two weeks are regular outliers: 3/12/2003-3/18/2003 and 12/17/2003-12/23 2003. This is consistent with the previous observation that ETMS data is incomplete for these weeks. Therefore, these two weeks were eliminated from further analysis. Next, the average percentage of additional flights was tabulated (Actual) and compared with the number of additional flights computed using the methodology presented herein. The results are shown in Table A.6.

Table A.6. Average Percentage of Additional Flights for Non-US North American Airports, 2003

Airport Name	Average % Additional Flights	
	Actual	Calc.
TJSJ	12.7%	7.5%
MMUN	31.7%	9.0%
CYYZ	8.6%	6.4%
CYYC	11.0%	7.6%
CYVR	7.7%	7.1%
MDSD	6.1%	11.6%
CYWG	13.1%	9.1%
MMMX	-1.4%	6.5%
CYOW	8.1%	8.5%

As can be seen, three airports (shaded) have a lower percentage of actual additional flights than what would be computed using the methodology presented herein. It was assumed that these airports were not cargo hubs and that the recommended methodology would be applied to them directly. For the other six airports, the Actual Average % Additional Flights factor was used.

A.2.2.2. European Airports

Next, the European airports with more than 166 weekly departures were analyzed. As mentioned in Section 6, a source of European data was available for validation. Unfortunately, data were not available for non-North American airports outside of Europe. Consequently, the Asian and South American airports in Table A.4 had to be dropped from the cargo analysis.

The candidate cargo hubs were analyzed for six weeks throughout the year 2003. The airports included in the analysis were: EGSS (Stansted, London), LFPG (Charles-De-Gaulle, Paris), EDDK (Cologne-Bonn, Germany), EDDL (Dusseldorf, Germany), EGNX (East Midlands, UK), EDDF (Frankfurt, Germany), EHAM (Amsterdam, Netherlands), EDDM (Munich, Germany), EBBR (Brussels, Belgium), and LIMC (Milan, Italy). Table A.7 displays the results.

Table A.7. Average Percentage of Additional Flights for European Airports, 2003

Airport Name	Average % Additional Flights	
	Actual	Calc.
EGSS	13.1%	7.1%
LFPG	9.4%	5.8%
EDDK	27.4%	7.4%

EDDL	7.6%	7.1%
EGNX	41.9%	8.9%
EDDF	6.4%	6.0%
EHAM	9.4%	6.4%
EDDM	4.4%	6.4%
EBBR	20.3%	7.0%
LIMC	9.2%	6.9%

As was the case in North America, the factor will only be applied to airports in which the actual number of departures is greater than what would be approximated from the methodology presented herein. It was assumed that these airports were not cargo hubs and that the recommended methodology would be applied to them directly. These airports in Europe are: EGSS, LFPG, EDDK, EDDL, EGNX, EDDF, EHAM, EBBR, and LIMC.

A.2.3. Final Approximation for Cargo Airports

In conclusion, Table A.8 shows the airports throughout the world that have a high number of cargo flights, and the factors that should be applied to estimate the number of actual flights.

Table A.8. Final Approximations for Cargo Airports

Airport Code	Airport Name	Additional Airports Factor
MEM	Memphis, TN	22.2%
IND	Indianapolis, IN	14.4%
SDF	Louisville, KY	38.5%
ANC	Anchorage, AK	55.0%
MIA	Miami, FL	14.7%
TJSJ	San Juan, PR	12.7%
MMUN	Cancun, Mexico	31.7%
CYYZ	Toronto, Canada	8.6%
CYYC	Calgary, Canada	11.0%
CYVR	Vancouver, Canada	7.7%
CYWG	Winnipeg, Canada	13.1%
EGSS	Stansted, London	13.1%
LFPG	Charles-De-Gaulle, Paris	9.4%

EDDK	Cologne-Bonn, Germany	27.4%
EDDL	Dusseldorf, Germany	7.6%
EGNX	East Midlands, UK	41.9%
EDDF	Frankfurt, Germany	6.4%
EHAM	Amsterdam, Netherlands	9.4%
EBBR	Brussels, Belgium	20.3%
LIMC	Milan, Italy	9.2%

A.3. Canceled Flights Outliers

Next, airports with an unusually large number of canceled flights were analyzed. First, the outliers for the 2000 were identified in Figures A.20 and A.21.

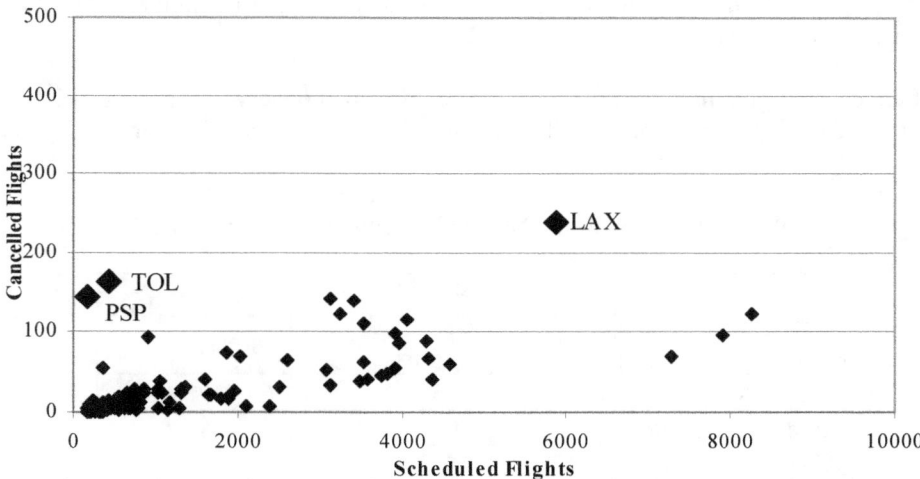

Figure A.20. May 2000 Scheduled vs. Canceled (Outliers)

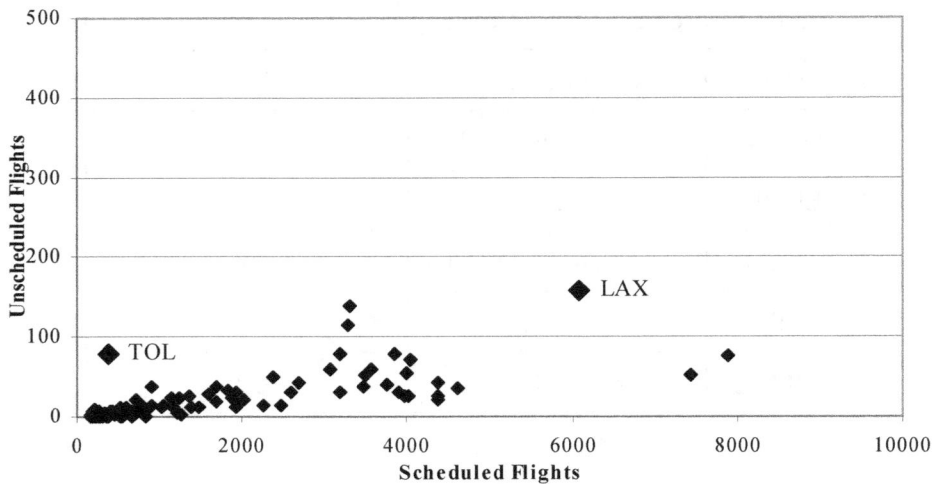

Figure A.21. October 2000 Scheduled vs. Canceled (Outliers)

Three outliers were identified: TOL (Toledo, Ohio), LAX (Los Angeles), and PSP (Palm Springs, CA) for May 2000. TOL operates as a "mini-hub" of BAX Global, a cargo company. A majority of the canceled flights in Toledo originate from this company, so it can be assumed that the high rate of cancellations is related to one particular company.

The number of canceled flights linked to specific carriers was also observed for LAX. Figures A.22 and A.23 present a table for May and October 2000, breaking down the canceled departures by carrier.

Table A.22. May 2000 Canceled Flights by Carrier at LAX

Carrier Name	Carrier Code	No. of Canceled Flights	% of Canceled Flights
United Airlines	UAL	105	26.05%
American Airlines	AAL	59	14.64%
Alaska Airlines	ASA	45	11.17%
Legend Airlines	LC	32	7.94%
Southwest Airlines	SWA	19	4.71%
Other	n/a	143	35.48%

Table A.23. October 2000 Canceled Flights by Carrier at LAX

Carrier Name	Carrier Code	No. of Canceled Flights	% of Canceled Flights
United Airlines	UAL	87	24.72%
Southwest Airlines	SWA	37	10.51%
Alaska Airlines	ASA	34	9.66%

American Airlines	AAL	28	7.95%
Air New Zealand	ANZ	10	2.84%
Other	n/a	156	44.32%

As can be seen, United Airlines (UAL) contributes to more canceled flights than any other carrier, about 25%. Upon further investigation, a good portion of the canceled UAL flights are daily flights which did not fly for the entire week. For instance, a flight from LAX to IAD was scheduled to depart at 3:30pm every day of the week. However, this flight did not operate once during the week, resulting in seven canceled flights. In the week observed for May 2000, 70 of the 105 canceled flights, and for October 2000, 49 of the 87 canceled flights were of this variety. This appears to be an obvious scheduling anomaly.

It is presumed that United was altering its operations at LAX after the flights had been scheduled. So, the outlying properties of TOL and LAX can be linked to specific carriers. And, as these two airports demonstrate a high number of cancellations across multiple weeks, they will be removed from further analysis.

The high number of cancellations at Palm Springs was not related to any particular carrier, but rather, was an overall issue with the airport. In Table A.24, the number of scheduled, flown, and canceled flights was broken down by flight date.

Table A.24. May 2000 Palm Springs (PSP) Analysis

	Flight Date						
	5/13/00	5/14/00	5/15/00	5/16/00	5/17/00	5/18/00	5/19/00
Scheduled	25	25	25	25	25	25	25
ETMS and OAG	0	1	2	1	1	2	20
OAG Only	25	24	23	24	24	23	5

As can be seen, nearly all of the flights for the week observed were in the OAG but not the ETMS data set. The issue can be traced back to an issue with the radar system at the Palm Springs airport. On May 7th, the radar system was completely shut down in lieu of repeated problems with the system [Duchin, 2000]. As ETMS data is based on radar data, any shutdown of a radar system will cause ETMS to miss actual flights. As flights slowly began to appear in ETMS over the week, it is assumed the radar system was restored. PSP will be removed for May 2000.

Taking a look at the results from 2001 reveals additional outliers.

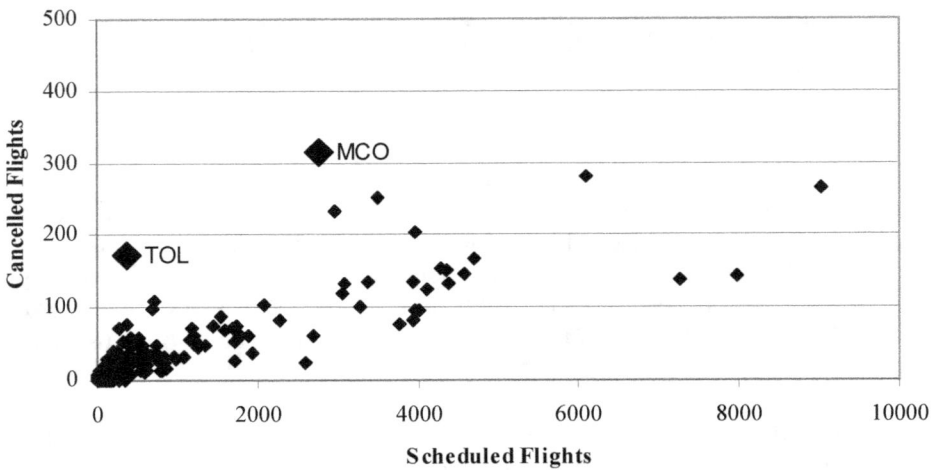

Figure A.22. May 2001 Scheduled vs. Canceled (Outliers)

There is one outlier that does not appear on the plot above: CVG (Cincinatti). The logic behind this decision was that it was such an extreme outlier that including it on the plot would require substantial compression of all the other losing a precision in the plot. Of the 4112 flights that were scheduled to depart from CVG, only 1986 flew. The issue at CVG is similar to the issue at MCO (Orlando). The two airports both serve as hubs for the carrier ComAir Inc., a sub-carrier of Delta Airlines. For the period of 2001 analyzed above, there was a pilot's strike at ComAir, resulting in a majority of the flights to be canceled [CNN, 2001]. Also, TOL is once again an outlier. An analysis of May and October 2003 reveal no additional outliers.

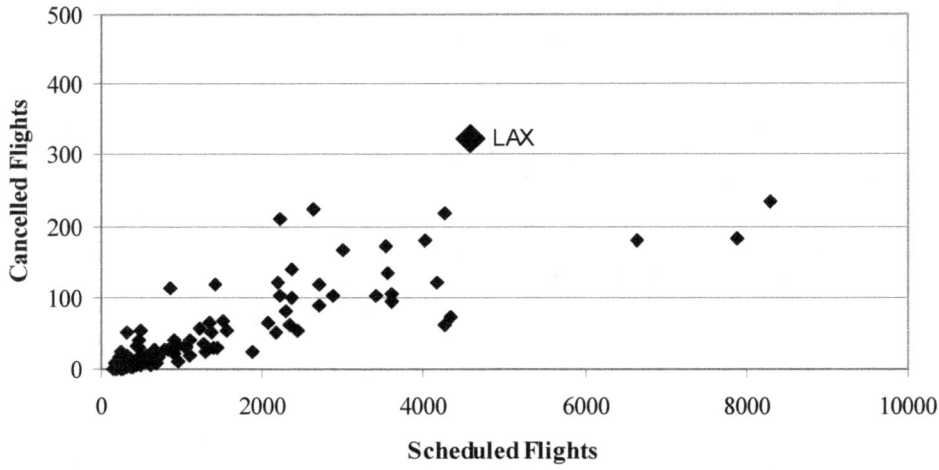

Figure A.23. May 2003 Scheduled vs. Canceled (Outliers)

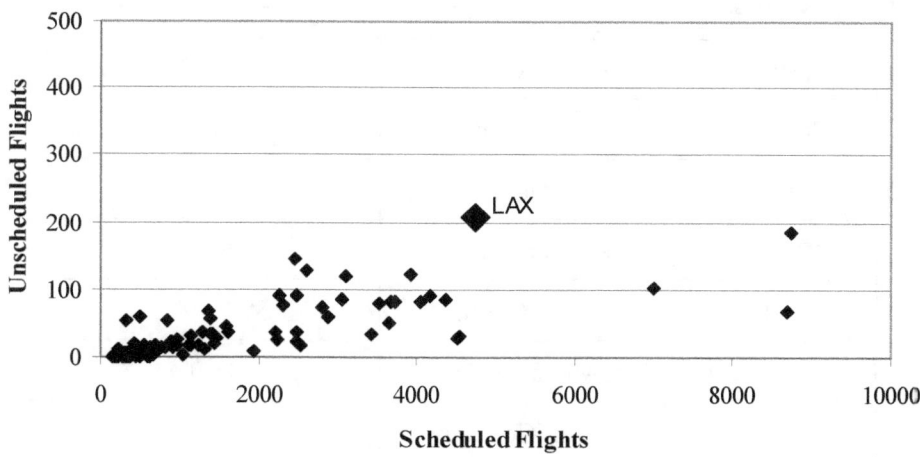

Figure A.24. October 2003 Scheduled vs. Canceled (Outliers)

In conclusion, the following airports will be removed from all data sets, as they demonstrated a high number of cancellations across multiple weeks: TOL and LAX. Additionally, PSP, MCO, and CVG demonstrated odd behavior for particular weeks only, so therefore PSP will only be eliminated from May 2000, and MCO and CVG will only be eliminated from May 2001.

For the airports listed as high-cargo volume airports, a special factor is recommended to be applied on an airport-by-airport basis. For cargo airports, the following scaling factors are recommended to be applied to OAG operations:

Table A.25. Final Approximations for Cargo Airports

Airport Code	Airport Name	Additional Airports Factor
MEM	Memphis, TN	22.2%
IND	Indianapolis, IN	14.4%
SDF	Louisville, KY	38.5%
ANC	Anchorage, AK	55.0%
MIA	Miami, FL	14.7%
TJSJ	San Juan, PR	12.7%
MMUN	Cancun, Mexico	31.7%
CYYZ	Toronto, Canada	8.6%
CYYC	Calgary, Canada	11.0%

CYVR	Vancouver, Canada	7.7%
CYWG	Winnipeg, Canada	13.1%
EGSS	Stansted, London	13.1%
LFPG	Charles-De-Gaulle, Paris	9.4%
EDDK	Cologne-Bonn, Germany	27.4%
EDDL	Dusseldorf, Germany	7.6%
EGNX	East Midlands, UK	41.9%
EDDF	Frankfurt, Germany	6.4%
EHAM	Amsterdam, Netherlands	9.4%
EBBR	Brussels, Belgium	20.3%
LIMC	Milan, Italy	9.2%

It is recommended that further work be undertaken to examine the possibility of developing a more robust methodology for adjusting for OAG operations at airports dominated by cargo operations and to expand coverage beyond North America and Europe.

Supplement B

B.1 Analysis of the Top 25 Busiest US Airports

A summary comparison of OAG scheduled flights with ETMS actual flights is presented in Tables B.1 through B.5 along with the number of flights computed after application of the recommended methodology.

Table B.1. Top 25 Busiest Airports May 2000

Airport Name	Sched Flights	Actual Flights	% Diff	Calc. Flights	% Diff
ORD	8266	8848	6.58%	8667	2.04%
ATL	7893	8233	4.13%	8284	-0.62%
DFW	7277	7713	5.65%	7649	0.83%
DTW	4577	4819	5.02%	4846	-0.56%
MSP	4376	4622	5.32%	4636	-0.30%
STL	4321	4553	5.10%	4578	-0.56%
PHX	4296	4504	4.62%	4552	-1.07%
EWR	4068	4406	7.67%	4313	2.10%
CVG	3970	4138	4.06%	4211	-1.76%

Table B.2. Top 25 Busiest Airports October 2000

Airport Name	Sched Flights	Actual Flights	% Diff	Calc. Flights	% Diff
ATL	7882	8354	5.65%	8272	0.98%
DFW	7435	8060	7.75%	7812	3.08%
DTW	4612	4870	5.30%	4882	-0.26%
MSP	4379	4726	7.34%	4639	1.84%
STL	4365	4758	8.26%	4624	2.81%
PHX	4360	4712	7.47%	4619	1.97%
EWR	4039	4515	10.54%	4283	5.14%
LGA	4004	4437	9.76%	4246	4.30%
BOS	3983	4432	10.13%	4224	4.68%

BOS	3915	4209	6.99%	4153	1.33%	CVG	3955	4190	5.61%	4195	-0.12%
IAH	3905	4131	5.47%	4143	-0.28%	PHL	3901	4260	8.43%	4138	2.85%
PHL	3824	4135	7.52%	4058	1.87%	IAH	3846	4111	6.45%	4081	0.73%
CLT	3738	3931	4.91%	3968	-0.93%	CLT	3759	3941	4.62%	3990	-1.23%
IAD	3572	3809	6.22%	3793	0.41%	DEN	3570	3905	8.58%	3791	2.91%
SEA	3539	3931	9.97%	3759	4.38%	SEA	3486	3830	8.98%	3703	3.31%
DEN	3533	3849	8.21%	3752	2.51%	IAD	3470	3697	6.14%	3686	0.29%
LGA	3496	3694	5.36%	3714	-0.53%	JFK	3299	3780	12.72%	3507	7.23%
SFO	3254	3547	8.26%	3459	2.47%	SFO	3190	3500	8.86%	3392	3.09%
JFK	3138	3548	11.56%	3337	5.94%	PIT	3172	3413	7.06%	3373	1.17%
PIT	3136	3321	5.57%	3335	-0.43%	LAS	3073	3642	15.62%	3269	10.24%
LAS	3087	3505	11.93%	3284	6.32%	MCO	2676	2981	10.23%	2851	4.37%
MCO	2607	2838	8.14%	2778	2.11%	BWI	2598	2761	5.90%	2769	-0.27%
BWI	2497	2660	6.13%	2662	-0.08%	DCA	2459	2692	8.66%	2622	2.60%
DCA	2375	2501	5.04%	2533	-1.29%	CLE	2267	2420	6.32%	2419	0.03%
CLE	2092	2232	6.27%	2234	-0.10%	RDU	2012	2166	7.11%	2150	0.75%

Table B.3. Top 25 Busiest Airports May 2001

Airport Name	Sched Flights	Actual Flights	% Diff	Calc. Flights	% Diff
ORD	9027	9419	4.16%	9447	-0.30%
ATL	7967	8272	3.69%	8360	-1.06%
DFW	7272	7491	2.92%	7644	-2.04%
DTW	4736	4857	2.49%	5012	-3.19%
STL	4565	4657	1.98%	4833	-3.79%
MSP	4384	4517	2.94%	4644	-2.82%
PHX	4382	4564	3.99%	4642	-1.71%
EWR	4277	4511	5.19%	4532	-0.47%
BOS	4138	4316	4.12%	4387	-1.64%
LGA	4027	4121	2.28%	4271	-3.63%
IAH	3994	4101	2.61%	4236	-3.29%
DEN	3992	4194	4.82%	4234	-0.95%
CLT	3939	4053	2.81%	4178	-3.09%
PHL	3933	4078	3.56%	4172	-2.31%
PIT	3768	3911	3.66%	3999	-2.25%
SEA	3422	3707	7.69%	3636	1.92%
IAD	3268	3386	3.48%	3474	-2.60%
LAS	3075	3325	7.52%	3271	1.62%
SFO	3064	3281	6.61%	3259	0.66%
JFK	2950	3225	8.53%	3139	2.65%
MCO	2760	2705	-2.03%	2939	-8.66%
BWI	2727	2833	3.74%	2905	-2.53%
DCA	2612	2726	4.18%	2783	-2.10%

CLE	2308	2333	1.07%	2463	-5.55%
RDU	2081	2334	10.84%	2464	-5.55%

Table B.4. Top 25 Busiest Airports May 2003

Airport Name	Sched Flights	Actual Flights	% Diff	Calc. Flights	% Diff
ORD	8311	8706	4.54%	8713	-0.08%
ATL	7917	8257	4.12%	8308	-0.62%
DFW	6650	6979	4.71%	7001	-0.32%
LAX	4582	4902	6.53%	4851	1.04%
DTW	4361	4462	2.26%	4620	-3.54%
CVG	4249	4406	3.56%	4503	-2.20%
PHX	4248	4385	3.12%	4502	-2.67%
MSP	4158	4309	3.50%	4408	-2.29%
IAH	4043	4209	3.94%	4287	-1.86%
CLT	3621	3794	4.56%	3845	-1.34%
DEN	3594	3731	3.67%	3816	-2.29%
EWR	3569	3829	6.79%	3790	1.01%
LGA	3554	3573	0.53%	3774	-5.64%
PHL	3427	3700	7.38%	3641	1.59%
SEA	3027	3131	3.32%	3220	-2.86%
LAS	2869	3166	9.38%	3054	3.53%
PIT	2739	2872	4.63%	2917	-1.57%
BOS	2727	2865	4.82%	2905	-1.38%
DCA	2433	2453	0.82%	2594	-5.77%
SFO	2369	2532	6.44%	2527	0.20%
IAD	2364	2562	7.73%	2522	1.58%
MCO	2341	2541	7.87%	2497	1.72%
JFK	2218	2520	11.98%	2367	6.05%
MDW	2213	2250	1.64%	2362	-4.99%
BWI	2182	2313	5.66%	2329	-0.71%

Table B.5. Top 25 Busiest Airports October 2003

Airport Name	Sched Flights	Actual Flights	% Diff	Calc. Flights	% Diff
ORD	8794	9152	3.91%	9209	-0.62%
ATL	8778	9129	3.84%	9192	-0.69%
DFW	7040	7320	3.83%	7405	-1.15%
CVG	4564	4758	4.08%	4832	-1.56%
DTW	4550	4653	2.21%	4818	-3.54%
MSP	4385	4507	2.71%	4645	-3.07%
PHX	4267	4382	2.62%	4522	-3.19%
IAH	4183	4273	2.11%	4434	-3.76%
DEN	3909	4098	4.61%	4147	-1.19%
CLT	3734	3935	5.11%	3963	-0.72%
PHL	3721	4018	7.39%	3950	1.70%
EWR	3708	3973	6.67%	3936	0.93%
STL	3580	3662	2.24%	3802	-3.82%
LGA	3549	3552	0.08%	3769	-6.12%
SEA	3180	3254	2.27%	3381	-3.92%
LAS	3046	3375	9.75%	3240	3.99%
BOS	2980	3116	4.36%	3171	-1.77%
PIT	2866	2970	3.50%	3051	-2.73%
IAD	2597	2794	7.05%	2767	0.95%
SFO	2527	2695	6.23%	2694	0.05%
DCA	2525	2563	1.48%	2692	-5.02%
MCO	2509	2660	5.68%	2675	-0.55%
JFK	2500	2743	8.86%	2665	2.84%
MDW	2330	2440	4.51%	2486	-1.87%
SLC	2267	2386	4.99%	2419	-1.39%

APPENDIX M : Fuel Burn and Emissions Module

This appendix provides details of the SAGE Version 1.5 FBE module in the form of pseudo-code [Lee 2005]. The module structure is shown below.

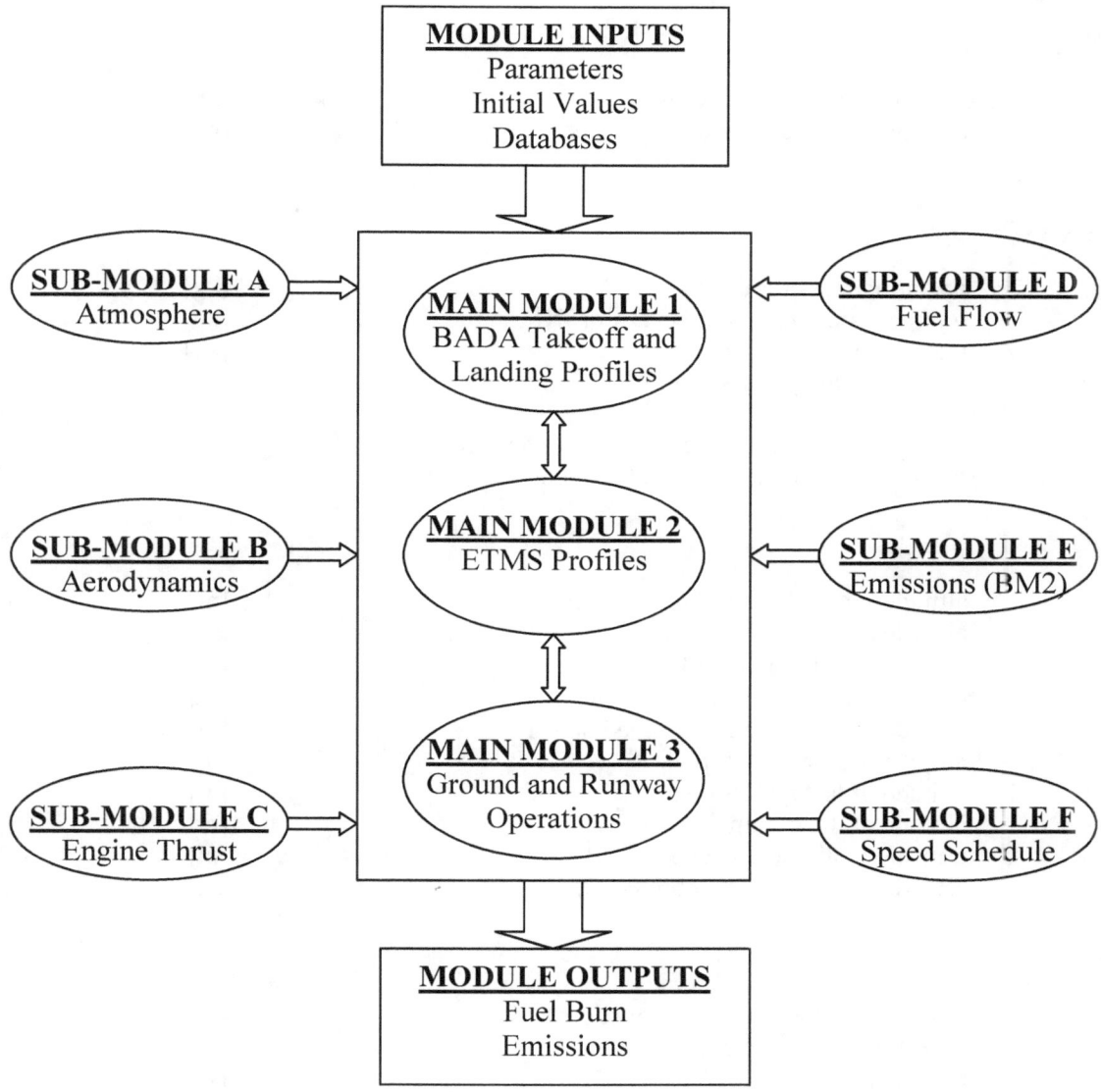

UNITS

BADA and INM use a mixture of both System International (SI) and English units in their equations and data files. These variations are taken into account through conversion factors applied directly within the equations. The result of this is that all outputs from the FBE module are expressed in SI units. Therefore, there is no need to modify BADA and INM data files prior to their use in the FBE module. This also applies to the ICAO and EDMS data files. All other input files (e.g., ETMS, OAG, etc.) must be converted to SI units before their use in the FBE module.

LIST OF PARAMETERS

(Note: The subscript i denotes i^{th} chord. Hence i+1 or i-1 appearing in the modules simply refers to $i+1^{th}$ or $i-1^{th}$ chord, respectively.)

Global

ROC_i	Rate of Climb for i^{th} chord	[m/s]
g	gravitational acceleration (9.81 m/s^2)	[m/s^2]
n	number of nodes	

Aircraft Mass

m_i	operating mass for i^{th} chord	[kg]
m_{ref}	reference initial mass in BADA or INM	[kg]
m_{max}	maximum takeoff mass	[kg]
m_{min}	operating empty mass	[kg]

Distance

B	takeoff distance coefficient in INM (averaged)	[ft/lb]
s_i	horizontal distance traversed for i^{th} chord	[m]

Altitude

h_i	altitude for i^{th} chord (midpoint of chord)	[m]
h_ih	altitude at head of ith chord	[m]
h_it	altitude at tail of ith chord	[m]
Δh_i	change in altitude for i^{th} chord	[m]
Δh_{calc}_i	altitude change calculated based on s_i	[m]
$h\_firstETMS$	altitude of first ETMS point	[m]
$h\_lastETMS$	altitude of last ETMS point	[m]
h_{trop}_i	altitude of the tropopause for i^{th} chord	[m]
h_{trans}	transition altitude for speed schedule	[m]
h_{des}	altitude for initial descent in BADA	[m]

Time

t_i	time for i^{th} chord (midpoint of chord)	[s]
t_ih	time at head of ith chord	[s]
t_it	time at tail of ith chord	[s]
Δt_i	change in time for i^{th} chord	[s]
$t_{taxiout}$	time to taxi-out in ASQP	[s]
t_{taxiin}	time to taxi-in ASQP	[s]

Speed

V_{GR}	speed at start of takeoff ground roll in INM	[knots]
V_{CAS}_i	calibrated airspeed for i^{th} chord	[m/s]

V_{GRS_i}	ground speed for i^{th} chord	[m/s]
V_{wind_i}	wind speed for i^{th} chord	[m/s]
V_i	true airspeed for i^{th} chord	[m/s]
V_ih	flight speed at head of ith chord	[m/s]
V_it	flight speed at tail of ith chord	[m/s]
ΔV_i	change in flight speed for i^{th} chord	[m/s]
M_i	Mach number for i^{th} chord, V_i / a_i	
V_{CAS_i}	calibrated airspeed for i^{th} chord (midpoint of chord)	[m/s]
V_{CAS_ih}	calibrated airspeed at head of ith chord	[m/s]
V_{CAS_it}	calibrated airspeed at tail of ith chord	[m/s]
V_{stall_TO}	stall speed in BADA (takeoff)	[knots]
V_{stall_LD}	stall speed in BADA (landing)	[knots]
$V_{cl\_1\_lo}$	1st standard climb V_{CAS} in BADA (low)	[knots]
$V_{cl\_1\_av}$	1st standard climb V_{CAS} in BADA (average)	[knots]
$V_{cl\_1\_hi}$	1st standard climb V_{CAS} in BADA (high)	[knots]
$V_{cl\_2\_lo}$	2nd standard climb V_{CAS} in BADA (low)	[knots]
$V_{cl\_2\_av}$	2nd standard climb V_{CAS} in BADA (average)	[knots]
$V_{cl\_2\_hi}$	2nd standard climb V_{CAS} in BADA (high)	[knots]
M_{cl_lo}	standard climb Mach number above h_{trans} in BADA (low)	
M_{cl_av}	standard climb Mach number above h_{trans} in BADA (average)	
M_{cl_hi}	standard climb Mach number above h_{trans} in BADA (high)	
M_{cr}	standard cruise Mach number in BADA	
$V_{des\_1\_lo}$	1st standard descent V_{CAS} in BADA (low)	[knots]
$V_{des\_1\_av}$	1st standard descent V_{CAS} in BADA (average)	[knots]
$V_{des\_1\_hi}$	1st standard descent V_{CAS} in BADA (high)	[knots]
$V_{des\_2\_lo}$	2nd standard descent V_{CAS} in BADA (low)	[knots]
$V_{des\_2\_av}$	2nd standard descent V_{CAS} in BADA (average)	[knots]
$V_{des\_2\_hi}$	2nd standard descent V_{CAS} in BADA (high)	[knots]
V_{STALL_CR}	stall speed in cruise in BADA	[knots]
V_{STALL_AP}	stall speed in approach in BADA	[knots]
M_{des_lo}	standard descent Mach number above h_{trans} in BADA (low)	
M_{des_av}	standard descent Mach number above h_{trans} in BADA (average)	
M_{des_hi}	standard descent Mach number above h_{trans} in BADA (high)	

Atmosphere

T_i	ambient temperature for i^{th} chord (midpoint of chord)	[K]
T_ih	ambient temperature at head of ith chord	[K]
T_it	ambient temperature at tail of ith chord	[K]
T_o_i	ambient temperature at sea level for i^{th} chord	[K]
T_{trop}	ambient temperature at tropopause	[K]
ϑ_i	temperature ratio for i^{th} chord (ambient to sea-level)	
P_i	ambient pressure for i^{th} chord (midpoint of chord)	[Pa]
P_ih	ambient pressure at head of ith chord	[Pa]
P_it	ambient pressure at tail of ith chord	[Pa]
P_o_i	ambient pressure at sea level for i^{th} chord	[Pa]
P_{trop}	ambient pressure at tropopause	[Pa]
P_v_i	saturation vapor pressure for i^{th} chord	[Pa]
δ_i	pressure ratio for i^{th} chord (ambient to sea-level)	[Pa]
δ_{trans}	pressure ratio (ambient to sea-level) at transition altitude	

ρ_i	ambient air density for i^{th} chord (midpoint of chord)	[kg/m³]
ρ_ih	ambient air density at head of ith chord	[kg/m³]
ρ_it	ambient air density at tail of ith chord	[kg/m³]
ρ_o_i	ambient air density at sea level for i^{th} chord	[kg/m³]
ρ_{trop}	ambient air density at tropopause	[kg/m³]
a_i	speed of sound for i^{th} chord (midpoint of chord)	[m/s]
a_ih	speed of sound at head of ith chord	[m/s]
a_it	speed of sound at tail of ith chord	[m/s]
a_o	speed of sound at ground level	[m/s]
γ	isentropic expansion coefficient for air (specific heat ratio)	
R	gas constant for air (287.04 J/kg-K)	[J/kg-K]
β_i	coefficient for saturation vapor pressure for i^{th} chord in BM2	
H_i	coefficient for $EINO_x$ for i^{th} chord in BM2	
$T\_taxiout$	ambient temperature for taxi-out	[K]
T_taxiin	ambient temperature for taxi-in	[K]
$P\_taxiout$	ambient pressure for taxi-out	[Pa]
P_taxiin	ambient pressure for taxi-in	[Pa]
$\theta\_taxiout$	temperature ratio for taxi-out (ambient to sea-level)	
θ_taxiin	temperature ratio for taxi-in (ambient to sea-level)	
$\delta\_taxiout$	pressure ratio for taxi-out (ambient to sea-level)	
δ_taxiin	pressure ratio for taxi-in (ambient to sea-level)	
$H\_taxiout$	coefficient for $EINO_x$ for taxi-out in BM2	
H_taxiin	coefficient for $EINO_x$ for taxi-in in BM2	
$P_v\_taxiout$	saturation vapor pressure for taxi-out	[Pa]
P_v_taxiin	saturation vapor pressure for taxi-in	[Pa]
$\beta\_taxiout$	coefficient for saturation vapor pressure for taxi-out in BM2	
β_taxiin	coefficient for saturation vapor pressure for taxi-in in BM2	

Aerodynamics

C_L_i	lift coefficient for i^{th} chord	
C_D_i	drag coefficient for i^{th} chord	
C_{DO_TO}	parasitic drag coefficient in BADA (takeoff)	
C_{D2_TO}	induced drag coefficient in BADA (takeoff)	
C_{DO_IC}	parasitic drag coefficient in BADA (initial climb)	
C_{D2_IC}	induced drag coefficient in BADA (initial climb)	
C_{DO_CR}	parasitic drag coefficient in BADA (cruise)	
C_{D2_CR}	induced drag coefficient in BADA (cruise)	
C_{DO_AP}	parasitic drag coefficient in BADA (approach)	
C_{D2_AP}	induced drag coefficient in BADA (approach)	
C_{DO_LD}	parasitic drag coefficient in BADA (landing)	
C_{D2_LD}	induced drag coefficient in BADA (landing)	
C_{DO_LDG}	parasite drag coefficient in BADA (landing gear)	
S	wing surface area	[m²]
L_i	lift force for i^{th} chord	[N]
D_i	drag force for i^{th} chord	[N]

Thrust

ESF_i	energy share factor for i^{th} chord

F_i	engine thrust (sum of all engines) for i^{th} chord	[N]
F_{max}_i	maximum takeoff and climb thrust for i^{th} chord	[N]
P_{max}	maximum power of turboprops	[W]
$C_{pow\_red\_i}$	reduced climb power coefficient for i^{th} chord	
C_{red}	maximum reduction in power to compute $C_{pow\_red\_i}$	
C_{red_jet}	maximum reduction in power for jets in BADA	
C_{red_turbo}	maximum reduction in power for turboprops in BADA	
C_{Tc_1}	1st maximum climb thrust coefficient in BADA	
C_{Tc_2}	2nd maximum climb thrust coefficient in BADA	
C_{Tc_3}	3rd maximum climb thrust coefficient in BADA	
C_{Tdes_high}	high altitude descent thrust coefficient in BADA	
C_{Tdes_low}	low altitude descent thrust coefficient in BADA	
C_{Tdes_AP}	approach thrust coefficient in BADA	
C_{Tdes_LD}	landing thrust coefficient in BADA	
E_{TO}	corrected net thrust at zero speed in INM (takeoff)	[lb]
E_{CL}	corrected net thrust at zero speed in INM (climb)	[lb]
$INMF_{TO}$	speed adjustment coefficient (takeoff) in INM	[lb/knot]
$INMF_{CL}$	speed adjustment coefficient (climb) in INM	[lb/knot]
G_{A_TO}	altitude adjustment coefficient (takeoff) in INM	[lb/ft]
G_{A_CL}	altitude adjustment coefficient (climb) in INM	[lb/ft]
G_{B_TO}	altitude-squared adjustment coefficient (takeoff) in INM	[lb/ft^2]
G_{B_CL}	altitude-squared adjustment coefficient (climb) in INM	[lb/ft^2]
$INMH_{TO}$	temperature adjustment coefficient (takeoff) in INM	[lb/K]
$INMH_{CL}$	temperature adjustment coefficient (climb) in INM	[lb/K]
$\%F_{oo}\_i$	percent power setting for i^{th} chord	
RO	Rated Output (single engine)	[N]
$INMRO$	INM-based Rated Output (single engine)	[N]
N	number of engines	
$\%F_{oo}\_taxiout$	percent power setting for taxi-out	
$\%F_{oo}\_taxiin$	percent power setting for taxi-in	

Fuel Flow and Fuel Burn

f_i	in-flight engine fuel flow for i^{th} chord	[kg/s]
C_{f1}	1st fuel flow coefficient in BADA	
C_{f2}	2nd fuel flow coefficient in BADA	
C_{f3}	1st descent fuel flow coefficient in BADA	
C_{f4}	2nd descent fuel flow coefficient in BADA	
C_{fcr}	cruise fuel flow correction coefficient in BADA	
FB_i	Fuel Burn for i^{th} chord	[kg]
TFB	Total Fuel Burn (sum of all chords)	[kg]
$FB\_taxiout$	Fuel Burn for taxi-out	[kg]
FB_taxiin	Fuel Burn for taxi-in	[kg]

Emissions

W_f_i	sea-level fuel flow for i^{th} chord, computed for BM2	[kg/s]
W_{f_TO}	sea-level fuel flow (ICAO takeoff)	[kg/s]
W_{f_CO}	sea-level fuel flow (ICAO climbout)	[kg/s]
W_{f_AP}	sea-level fuel flow (ICAO approach)	[kg/s]
W_{f_ID}	sea-level fuel flow (ICAO idle)	[kg/s]

W_{ff_i}	fuel flow factor for i^{th} chord, computed for BM2	[kg/s]
W_{ff_TO}	fuel flow factor	[kg/s]
W_{ff_CO}	fuel flow factor	[kg/s]
W_{ff_AP}	fuel flow factor	[kg/s]
W_{ff_ID}	fuel flow factor	[kg/s]
$REIHC\_i$	Reference Emission Index for HC for i^{th} chord	[g/kg-fuel]
$REICO\_i$	Reference Emission Index for CO for i^{th} chord	[g/kg-fuel]
$REINO_x\_i$	Reference Emission Index for NO_x for i^{th} chord	[g/kg-fuel]
$EIHC\_i$	Emission Index for HC for i^{th} chord	[g/kg-fuel]
$EICO\_i$	Emission Index for CO for i^{th} chord	[g/kg-fuel]
$EINO_x\_i$	Emission Index for NOx for i^{th} chord	[g/kg-fuel]
CO_2_i	amount of CO_2 emissions for i^{th} chord	[g]
H_2O_i	amount of H_2O emissions for i^{th} chord	[g]
SO_2_i	amount of SO_2 emissions for i^{th} chord	[g]
HC_i	amount of HC emissions for i^{th} chord	[g]
CO_i	amount of CO emissions for i^{th} chord	[g]
NO_x_i	amount of NO_x emissions for i^{th} chord	[g]
TCO_2	total amount of CO_2 emissions (sum of all chords)	[g]
THC	total amount of HC emissions (sum of all chords)	[g]
TCO	total amount of CO emissions (sum of all chords)	[g]
TNO_x	total amount of NO_x emissions (sum of all chords)	[g]
$CO_2\_taxiout$	amount of CO_2 emissions for taxi-out	[g]
CO_2_taxiin	amount of CO_2 emissions for taxi-in	[g]
$H_2O\_taxiout$	amount of H_2O emissions for taxi-out	[g]
H_2O_taxiin	amount of H_2O emissions for taxi-in	[g]
$SO_2\_taxiout$	amount of SO_2 emissions for taxi-out	[g]
SO_2_taxiin	amount of SO_2 emissions for taxi-in	[g]
$REIHC\_taxiout$	Reference Emission Index for HC for taxi-out	[g/kg-fuel]
$REIHC\_taxiin$	Reference Emission Index for HC for taxi-in	[g/kg-fuel]
$REICO\_taxiout$	Reference Emission Index for CO for taxi-out	[g/kg-fuel]
$REICO\_taxiin$	Reference Emission Index for CO for taxi-in	[g/kg-fuel]
$REINO_x\_taxiout$	Reference Emission Index for NO_x for taxi-out	[g/kg-fuel]
$REINO_x\_taxiin$	Reference Emission Index for NO_x for taxi-in	[g/kg-fuel]
$EIHC\_taxiout$	Emission Index for HC for taxi-out	[g/kg-fuel]
$EIHC\_taxiin$	Emission Index for HC for taxi-in	[g/kg-fuel]
$EICO\_taxiout$	Emission Index for CO for taxi-out	[g/kg-fuel]
$EICO\_taxiin$	Emission Index for CO for taxi-in	[g/kg-fuel]
$EINO_x\_taxiout$	Emission Index for NOx for taxi-out	[g/kg-fuel]
$EINO_x\_taxiin$	Emission Index for NOx for taxi-in	[g/kg-fuel]
$HC\_taxiout$	amount of HC emissions for taxi-out	[g]
HC_taxiin	amount of HC emissions for taxi-in	[g]
$CO\_taxiout$	amount of CO emissions for taxi-out	[g]
CO_taxiin	amount of CO emissions for taxi-in	[g]
$NO_x\_taxiout$	amount of NO_x emissions for taxi-out	[g]
NO_x_taxiin	amount of NO_x emissions for taxi-in	[g]
$W_f\_taxiout$	sea-level fuel flow for taxi-out (ICAO idle)	[kg/s]
W_f_taxiin	sea-level fuel flow for taxi-in (ICAO idle)	[kg/s]
$W_{ff}\_taxiout$	fuel flow factor for taxi-out	[kg/s]
W_{ff}_taxiin	fuel flow factor for taxi-in	[kg/s]

ASSUMPTIONS

Atmosphere

ISA assumed
No wind assumed ($V_{wind\_i} = 0$)
Relative humidity of 60% assumed
Specific heat ratio of 1.4 assumed ($\gamma = 1.4$)

Trajectory

Straight line between data points assumed
Rule of midpoint used (see Supplement 1)

Operability

BADA nominal takeoff mass assumed, however, takeoff mass based on load factor and stage length (such as the ones in INM) should be used whenever available.

Performance

The fundamental energy equation assumed for aircraft maneuvering
INM procedures assumed for LTO performance
INM engine thrust model assumed for takeoff and landing
BADA aerodynamic model assumed for drag coefficients
BM2 assumed for HC, CO, and NO_x emissions
Installation effects of engine air bleed corrected by the coefficients given in BM2

MAIN MODULE 1: TAKEOFF AND LANDING TRAJECTORIES

1. Generate altitude nodes for takeoff and landing

   a. For takeoff, create nodes at 0 m, 0 m, 304.80 m (1000 ft), 457.20 m (1500 ft), 609.60 m (2000 ft), 914.40 m (3000 ft), and 1219.20 m (4000 ft). Above 1219.20 m (4000 ft), create nodes at every 609.60 m (2000 ft) up to the closest location to the second ETMS point (or the first OAG cruise point). Then make the second ETMS point (or the first OAG cruise point) the last point of the takeoff profile. Use the following loop algorithm.

Note: At least up to 4000 ft (6th chord), a takeoff profile should be generated and used solely even if ETMS data points exist at lower altitudes.

    For n nodes and i chords for takeoff
        $h\_1h = 0$
        $h\_1t = 0$
    if <= 7 node, $h\_i+1h = h\_it$ where:
        $h\_2t = 304.80$
        $h\_3t = 457.20$
        $h\_4t = 609.60$
        $h\_5t = 914.40$
        $h\_6t = 1219.20$

if > 7 node, $h\_it = h\_ih + 609.60$ and $h\_i+1h = h\_it$
if n node, $h\_it = h\_secondETMS$ (or $h\_firstOAG$)

b. For landing, create nodes at 0 m, 304.80 m (1000 ft), 457.20 m (1500 ft), and 914.40 m (3000 ft). Above 914.40 m (3000 ft), create nodes at every 609.60 m (2000 ft) up to the closest location to the last ETMS point (or the last OAG cruise point). Then make the last ETMS point (or the last OAG cruise point) the starting point of landing. Use the following loop algorithm.

For n nodes and i chords for landing
$\quad h\_1h = h\_lastETMS$ (or $h\_lastOAG$)
if $< n$ and > 914.40, $h\_it = h\_ih - 609.60$ and $h\_i+1h = h\_it$
if $< n$ and ≤ 914.40, $h\_i+1h = h\_it$ where:
$\quad h\_i-3t = 914.40$
$\quad h\_i-2t = 457.20$
$\quad h\_i-1t = 304.80$
if n node, $h\_it = 0$

c. Combine the created points with ETMS data points, and number all of them in sequence.

d. Compute Δh_i and h_i for each chord.

$\Delta h\_i = h\_it - h\_ih$
$h\_i = (h\_ih + h\_it) / 2$

Note1: At this point, a complete flight position profile of BADA-ETMS-BADA must have been created.

Note2: The steps below are only necessary for the BADA created profile.

2. Determine atmospheric conditions for a given chord

   a. Compute ISA T_i, P_i, ρ_i, and a_i for each chord (see Sub-module A)
   b. Compute ISA T_ih, P_ih, ρ_ih, and a_ih at head of each chord as well as ISA T_it, P_it, ρ_it, and a_it at tail of each chord.
   c. Compute temperature and pressure ratios. Use the following loop algorithm.

Note1: Do not attempt to compute the atmospheric conditions and speed schedule for all chords of the entire profile in one step. Rather, the atmospheric conditions and speed schedule must be generated chord-by-chord by repeating steps 2 through 7. This is because the speed schedule depends on aircraft mass, which is debited across each chord.

Note2: "Temperature," "Pressure," "Density," and "Speed of Sound" refer to their functional forms given in Sub-module A.

For n nodes

$T\_1h = T_o$
If $h\_it < h_{trop}$, $T_it = $ Temperature (h_it)
If $h\_it > h_{trop}$, $T\_it = 216.65$
If $< n$ node, $T\_i+1h = T\_it$

P_1h = P_o
If h_it < h_{trop}, P_it = Pressure (T_it)
If h_it > h_{trop}, P_it = Pressure (h_it)
If < n node, P_i+1h = P_it

ρ_1h = ρ_o
If h_it < h_{trop}, ρ_it = Density (T_it)
If h_it > h_{trop}, ρ_it = Density (h_it)
If < n node, ρ_i+1h = ρ_it

a_1h = a_o
If h_it < h_{trop}, a_it = Speed of Sound (T_it)
If h_it > h_{trop}, a_it = 295.07
If < n node, a_i+1h = a_it

$$\boxed{\theta\_i = \frac{T\_i}{288.15}} \text{ and } \boxed{\delta\_i = \frac{P\_i}{101325}}$$

3. Determine transition altitude (h_{trans}), mass (m_1), and speed (V_1h) (see Sub-module F-1) and compute speed schedule for a given chord (see Sub-module F-2)

4. Determine aerodynamic coefficients (C_L_i and C_D_i) and lift (L_i) and drag (D_i) forces for each chord (see Sub-module B)

5. Compute thrust (F_i) for each chord from BADA engine thrust model and compute Δt_i for each chord (see Sub-module C-2)

6. Compute fuel flow (f_i) and fuel burn (FB_i) for each chord, and debit aircraft weight (see Sub-module D)

7. If takeoff ground roll (first chord), climb, or descent, compute the ground track distance (s_i) and go to step 8. If acceleration, compute the s_i and height gain (Δh_{calc}_i). Compare Δh_{calc}_i with estimated height gain (Δh_i). Adjust the Δh_i and repeat steps 2 through 7. Use the following algorithm.

Use the INM FLAPS file for B values.

In takeoff mode (ROC>0):

If i = 1,

$$\boxed{s\_i = \frac{B \cdot \theta\_i}{4.2509 \times 3.2808} \frac{\left(m\_i \cdot g / \delta\_i\right)^2}{\left(F\_i / \delta\_i\right)}}$$ and go to step 8

If i = 3, 4, or 6, or join chord

$$s\_i = \frac{(V\_it)^2 - (V\_ih)^2}{2g} \left[\frac{(F\_i/\delta\_i)}{(m\_i \cdot g/\delta\_i)} - \frac{D\_i}{L\_i} - \frac{ROC\_i}{V\_i} \right]^{-1}$$ and $$\Delta h_{calc}\_i = s\_i \cdot \frac{ROC\_i}{V\_i}$$. Then check the following:

If $|\Delta h\_i - \Delta h_{calc}\_i| \leq 1$, proceed to step 8. Otherwise,
   If $\Delta h > \Delta h_{calc} + 1$, then
      decrease h_it by $|\Delta h_{calc} - \Delta h|/2$, update h_i+1h (= h_it), and repeat steps 2 through 7
   If $\Delta h < \Delta h_{calc} + 1$, then
      increase h_it by $|\Delta h_{calc} - \Delta h|/2$, update h_i+1h (= h_it), and repeat steps 2 through 7

Note: This is the iteration loop required for acceleration chords in order for the altitudes to converge to the final values.

If $V_{CAS}\_it - V_{CAS}\_ih \leq 0$, or all other cases wehere ROC_i > 0 in takeoff

$$s\_i = \sqrt{\left(\frac{\Delta h\_i}{ROC\_i/V\_i}\right)^2 - (\Delta h\_i)^2}$$ and go to step 8

If landing mode (ROC_i < 0):

$$s\_i = \frac{\Delta h\_i}{\tan(-3°)}$$ for jet aircraft

$$s\_i = \frac{\Delta h\_i}{\tan(-5°)}$$ for turboprop aircraft

and go to step 8

8. Move to the next chord (i = i +1). Repeat steps 2 through 7 for every chord of the entire takeoff or landing profile.

Note: If the takeoff profile is complete, proceed to step 9. When the landing profile is complete, proceed to step 10.

9. Go to main module 2 step 1

10. Determine emissions for each chord of entire takeoff and landing profiles (see Sub-module E)

    a. Compute CO_2_i directly from fuel burn
    b. Compute $EIHC\_i$, $EICO\_i$, and $EINO_x\_i$ using BM2
    c. Compute HC_i, CO_i, and NO_x_i

11. Compute total fuel burn (TFB) and total emissions (TCO$_2$, THC, TCO, and TNO$_x$) for entire takeoff and landing profiles

MAIN MODULE 2: CRUISE TRAJECTORY

1. Preprocess data based on chorded ETMS profiles

    b. Compute Δh_i, Δt_i, and ROC_i ($\Delta h\_i/\Delta t\_i$) for each chord (see Figure 1)
    c. Eliminate non-physical points

        i. Make sure ROC_i < 25 m/s
        ii. Make sure no reversing in direction of climb or descent
        iii. Preserve Δt_i even if any data point gets dropped

2.1. Compute ROC again on preprocessed data

    a. Compute Δh_i, Δt_i, and ROC_i for each chord
    b. Compute h_i, ΔV_i, and V_i for each chord (see Figure 1)

2.2. Compute ROC for a delay chord with the following pre-determined data:

Altitude at head (h_ih), altitude at tail (h_it), head speed (V$_{GRS}$_ih), tail speed (V$_{GRS}$_it), and Δt_i (set this to 0.0001 if there is no delay)

    a. Compute Δh_i and ROC_i
    b. Compute h_i, ΔV_i, and V_i
    c. Get the aircraft mass from the previous ETMS chord (m_i = m_i-1 – FB_i-1). Then pass the fuel burn debited aircraft mass to the next ETMS chord (m_i+1 = m_i – FB_i) after FB_i is computed for the delay chord.

Note1: All speed input to SAGE must be true airspeed (referred to as V). However, ETMS speed is only ground speed (therefore, this will now be called V$_{GRS}$). In order to convert V$_{GRS}$ to V, account for wind speed as follows:

$$V = V_{GRS} + V_{wind}$$

Note2: Headwind is denoted by a positive value of V$_{wind}$ whereas tail wind is denoted by a negative value of V$_{wind}$. Initially, we set V$_{wind}$ = 0 and should be able to update it whenever wind data become available.

Note3: Any speed we generate from BADA is already in true airspeed. Therefore, there is no need for conversion.

3. Determine atmospheric conditions for all chords before proceeding to 4 (see Sub-module A)

    a. Compute ISA T_i, P_i, ρ_i, and a_i for each chord
    b. Compute M_i (V_i /a_i) for each chord

4. Determine aerodynamic coefficients (C$_L$_i and C$_D$_i) and lift (L_i) and drag (D_i) forces for each chord (see Sub-module B)

5. Compute thrust (F_i) for each chord from the fundamental energy equation (see Sub-module C-1)

6. Compute fuel flow (f_i) and fuel burn (FB_i) for each chord, and debit aircraft weight (see Sub-module D)

7. Determine emissions for each chord (see Sub-module E)

   a. Compute CO_2_i directly from fuel burn
   b. Compute $EIHC\_i$, $EICO\_i$, and $EINO_x\_i$ using BM2
   c. Compute HC_i, CO_i, and NO_x_i

8. Repeat steps 4 through 7 for every chord of the entire profile and compute total fuel burn (TFB) and total emissions (TCO_2, THC, TCO, and TNO_x) for entire ETMS profile.

9. Go to main module 1 step 3

MAIN MODULE 3: GROUND OPERATIONS

1. Compute fuel burn based on 7% idle power setting and corrected fuel flow (see Sub-module E-2 step 2)
Note: Use ASQP for $t_{taxiout}$ and t_{taxiin}.

$$FB\_taxiout = W_{ff}\_ID \times t_{taxiout}$$ and $$FB\_taxiin = W_{ff}\_ID \times t_{taxiin}$$

2. Mass balance to compute CO_2, H_2O, and SO_2 emissions from fuel burn (see Sub-module E-1)

3. Set ground temperature and pressure
Note: The subscript n denotes the last chord of the flight profile.

$$T\_taxiout = T\_1h$$ and $$T\_taxiin = T\_nt$$

$$P\_taxiout = P\_1h$$ and $$P\_taxiin = P\_nt$$

4. BM2 to compute HC, CO, and NO_x emissions

   a. Set percent power setting of engine thrust

   $$\%F_{oo}\_taxiout = 0.07$$

   b. Determine $W_f\_taxiout$ from ICAO W_f vs. $\%F_{oo}$ curve constructed in Sub-module E-2 step 1

   c. Compute temperature and pressure ratios

   $$\theta\_taxiout = \frac{T\_taxiout}{288.15}$$ and $$\delta\_taxiout = \frac{P\_taxiout}{101325}$$

   d. Compute $W_{ff}\_taxiout$, which corrects $W_f\_$ taxiout for altitude

$$W_{ff\_taxiout} = \frac{W_{f\_taxiout}}{\delta\_taxiout} \theta\_taxiout^{3.8}$$

    e. Determine REIHC_taxiout, REICO_taxiout, and REINOx_taxiout from REIHC vs. W_{ff}, REICO vs. W_{ff}, and REINOx vs. W_{ff} curves constructed in Sub-module E-2 step 2

    f. Replace taxiout with taxiin and repeat steps a-e above

5. Compute EIHC_taxiout, EICO_taxiout, and EINOx_taxiout as well as EIHC_taxiin, EICO_taxiin, and EINOx_taxiin (see Sub-module E-2 step 3)

6. Compute amount of emissions for taxi-out and taxi-in (see Sub-module E-2 step 4)

Note: Fuel burn and emissions for taxi-out must be computed before MAIN MODULE 1 AND those for taxi-in must be computed after MAIN MODULE 1 when all chords (taxi-out, takeoff, cruise, landing, and taxi-in chords) are connected.

SUB-MODULE A: ATMOSPHERIC CONDITIONS

1. Determine tropopause altitude [m]

$$h_{trop\_i} = 11000 + \frac{1000 \times (T_o\_i - 288.15)}{6.5}$$

For ISA conditions, the tropopause is at 11000 m altitude.

2. Determine ambient temperature [K]

Temperature at sea level

$$T_o\_i = 288.15$$

Below the tropopause ($h\_i < h_{trop}\_i$), the temperature is calculated as a function of altitude as follows:

$$T\_i = T_o\_i - 0.0065 \times h\_i$$

Above the tropopause ($h\_i > h_{trop}\_i$), the temperature is constant.

$$T\_i = T_{trop} = 216.65$$

3. Determine ambient pressure [Pa]

Pressure at sea level

$$P_o\_i = 101325$$

Below the tropopause ($h\_i < h_{trop}\_i$), the pressure at altitude can be determined from the temperature at altitude by the following:

$$P\_i = P_o\_i \left(\frac{T\_i}{T_o\_i} \right)^{5.2579}$$

Above the tropopause ($h\_i > h_{trop}\_i$), the following is used:

$$P\_i = P_{trop} \cdot e^{-\left(\frac{g}{RT_{trop}}\right)(h\_i - h_{trop}\_i)} \quad \text{where} \quad P_{trop} = 22619$$

4. Determine ambient air density [kg/m³]

Air density at sea level

$$\rho_o\_i = 1.225 \left(\frac{288.15}{T_o\_i} \right)$$

Below the tropopause ($h\_i < h_{trop}\_i$), the air density is calculated as a function of temperature as follows:

$$\rho\_i = \rho_o\_i \left(\frac{T\_i}{T_o\_i} \right)^{4.2586}$$

Above the tropopause ($h\_i > h_{trop}\_i$), the air density is calculated as follows:

$$\rho\_i = \rho_{trop} \cdot e^{-\left(\frac{g}{RT_{trop}}\right)(h\_i - h_{trop}\_i)} \quad \text{where} \quad \rho_{trop} = 0.36364$$

5. Determine speed of sound [m/s]

Speed of sound at sea level

$$a_o = 340.29$$

Below the tropopause ($h < h_{trop}$), the speed of sound is calculated as a function of temperature.

$$a\_i = a_o \sqrt{\frac{T\_i}{288.15}}$$

Above the tropopause ($h > h_{trop}$), the speed of sound is constant.

$$a_{trop} = 295.07$$

6. Determine Mach number

$$M\_i = \frac{V\_i}{a\_i}$$

SUB-MODULE B: AERODYNAMIC DRAG

1. Compute lift coefficient

$$C_{L\_}i = \frac{2 \cdot m\_i \cdot g}{\rho\_i \cdot V\_i^2 \cdot S}$$

2. Compute drag coefficient

a. Takeoff (If i = 1, 2, or 3)

$$C_{D\_}i = C_{D0\_TO} + C_{D2\_TO} \cdot (C_{L\_}i)^2$$

b. Acceleration in initial climb (If i = 4)

$$C_{D\_}i = C_{D0\_IC} + C_{D2\_IC} \cdot (C_{L\_}i)^2$$

c. Climb out / cruise / initial descent (If i ≥ 5)

$$C_{D\_}i = C_{D0\_CR} + C_{D2\_CR} \cdot (C_{L\_}i)^2$$

d. Approach (If < 2438 m (8,000 ft) after cruise segments)

$$C_{D\_}i = C_{D0\_AP} + C_{D2\_AP} \cdot (C_{L\_}i)^2$$

e. Landing (If < 914 m (3,000 ft) after cruise segments)

$$C_{D\_}i = C_{D0\_LD} + C_{D0\_LDG} + C_{D2\_LD} \cdot (C_{L\_}i)^2$$

In case C_{D0_TO}, C_{D2_TO}, C_{D0_IC}, C_{D2_IC}, C_{D0_AP}, C_{D2_AP}, C_{D0_LD}, C_{D2_LD} and C_{D0_LDG} are zero, do not use the cruise coefficients as prescribed from BADA. Rather, use the following scale factors to convert the cruise coefficients to the appropriate mode:

Aircraft Category	Initial Climb		Takeoff		Approach		Landing		Landing Gear
	Parasitic	Induced	Parasitic	Induced	Parasitic	Induced	Parasitic	Induced	Parasitic
Jets	1.143	1.071	1.476	1.010	1.957	0.992	3.601	0.932	1.037
Turboprops	1.000	1.000	1.220	0.948	1.279	0.940	1.828	0.916	0.496

3. Compute lift force [N]

$$L\_i = \frac{C_{L\_}i \cdot \rho\_i \cdot V\_i^2 \cdot S}{2}$$

4. Compute drag force [N]

$$D\_i = \frac{C_{D}\_i \cdot \rho\_i \cdot V\_i^2 \cdot S}{2}$$

SUB-MODULE C: DETERMINATION OF ENGINE THRUST AND RATE OF CLIMB

Contribution from all engines is determined. When trajectory data is available, use C-1. Otherwise, use C-2. The relationship between engine thrust and rate of climb is established by employing the fundamental energy equation below.

$$(F - D) \cdot V = mg \frac{\Delta h}{\Delta t} + mV \frac{\Delta V}{\Delta t} \quad \text{where} \quad \frac{\Delta h}{\Delta t} = ROC$$

C-1. Thrust based on fundamental energy equation and ETMS rate of climb

1. Determine F [N]

If ROC≠0, $$F\_i = \frac{m\_i \cdot g}{V\_i}\left(ROC\_i + \frac{V\_i}{g} \cdot \frac{\Delta V\_i}{\Delta t\_i}\right) + D\_i$$

If ROC=0, $F\_i = D\_i$

C-2. BADA engine thrust and determination of rate of climb

1. Determine engine thrust [N] (contribution from all engines)

a. Compute maximum takeoff thrust first

For jet aircraft:

$$INMRO = 4.4482 E_{TO}$$

$$F_{max}\_i = 4.4482 \times N\left(E_{TO} + \frac{INMF_{TO}}{0.51444} V_{CAS}\_i + G_{A\_TO}(3.2808 \times h\_i) + G_{B\_TO}(3.2808 \times h\_i)^2 + INMH_{TO} \cdot T\_i\right)\delta\_i$$

For turboprop aircraft:
$$INMRO = 4.4482 E_{TO}$$

If h_i < 914 m (3000 ft)
$$F_{max}\_i = 4.4482N \times E_{TO}$$

If h_i ≥ 914 m (3000 ft)
$$F_{max}\_i = 0.5966 \times 4.4482N \times E_{TO}$$

b. Takeoff mode

For jet aircraft:

If i = 1, 2, 3, or 4, $\boxed{F\_i = F_{max}\_i}$

Otherwise,

$$\boxed{F\_i = 4.4482 \times N\left(E_{CL} + \frac{INMF_{CL}}{0.51444} V_{CAS}\_i + G_{A\_CL}(3.2808 \times h\_i) + G_{B\_CL}(3.2808 \times h\_i)^2 + INMH_{CL} \cdot T\_i\right)\delta\_i}$$

For turboprop aircraft:

$\boxed{F\_i = F_{max}\_i}$

c. Landing mode

Check the BADA thrust coefficients first as following:

If $C_{Tdes\_high} < 0$, $C_{Tdes\_high} = 0.01$
If $C_{Tdes\_low} < 0$, $C_{Tdes\_low} = 0.01$
If $C_{Tdes\_AP} \leq 0$, $C_{Tdes\_AP} = C_{Tdes\_low}$
If $C_{Tdes\_LD} \leq 0$, $C_{Tdes\_LD} = C_{Tdes\_low}$

Then,

if $h \geq h_{des}$: $\boxed{F\_i = C_{Tdes\_high} \cdot F_{max}\_i}$
if 2438 m (8,000 ft) $\leq h < h_{des}$: $\boxed{F\_i = C_{Tdes\_low} \cdot F_{max}\_i}$
if 914 m $\leq h <$ 2438 m (8,000 ft), and $V_{CAS}\_i < 0.51444 V_{STALL\_CR}$: $\boxed{F\_i = C_{Tdes\_AP} \cdot F_{max}\_i}$
if $h <$ 914 m (3,000 ft), and $V_{CAS}\_i < 0.51444 V_{STALL\_AP}$: $\boxed{F\_i = C_{Tdes\_LD} \cdot F_{max}\_i}$

Use BADA OPF files for values of V_{STALL_CR} and V_{STALL_AP}.

2. Determine Energy Share Factor (ESF_i) for each chord

If i = 1, $\boxed{ESF\_i = 0}$.

If i = 3, 4, or 6, $\boxed{ESF\_i = 0.3}$.

If i is 1) joining chord between ETMS trajectory and takeoff (or landing) profile, and 2) deceleration ($\Delta V\_i < 0$) in takeoff or acceleration ($\Delta V\_i > 0$) in descent, $\boxed{ESF\_i = 1.7}$

Otherwise, $\boxed{ESF\_i = \left[1 + \left(\frac{V\_i}{g}\right)\left(\frac{\Delta V\_i}{\Delta h\_i}\right)\right]^{-1}}$

3. Determine C_{pow_red} for each chord

$$C_{pow\_red}\_i = 1 - C_{red} \frac{m_{max} - m\_i}{m_{max} - m_{min}}$$

For jet aircraft, and if i ≥ 3 for takeoff chords,

$$C_{red} = C_{red\_jet} = 0.15$$

For turboprop aircraft, and if i ≥ 3 for takeoff chords,

$$C_{red} = C_{red\_turbo} = 0.25$$

For piston engine, or for all other cases,

$$C_{red} = 0$$

4. Compute ROC [m/s] for each chord

$$ROC\_i = \frac{(F\_i - D\_i) \cdot V\_i \cdot C_{pow\_red}\_i \cdot ESF\_i}{m\_i \cdot g}$$

If (i ≠ 3, 4, or 6) and ((ROC_i < 2 m/s or F_i < D_i during takeoff) or (ROC_i > 25 m/s)), set ROC as following:

    ROC_i = 12 m/s if h_i < 3047.9 m (10000 ft)
    ROC_i = 8 m/s if 3047.9 m ≤ h_i < 4571.8 m (15000 ft)
    ROC_i = 5 m/s if 4571.8 m ≤ h_i < 6095.7 m (20000 ft)
    ROC_i = 3 m/s if 6095.7 m ≤ h_i < 8229.2 m (27000 ft)
    ROC_i = 2 m/s if h_i ≥ 8229.2 m

Also, compute F_i again using the following equations, update F_i, and proceed to main module 1 step 6.

$$\Delta t\_i = \frac{\Delta h\_i}{ROC\_i}$$

$$F\_i = \frac{m\_i \cdot g}{V\_i}\left(ROC\_i + \frac{V\_i}{g} \cdot \frac{\Delta V\_i}{\Delta t\_i}\right) + D\_i$$

If approach mode,

For jets:

$$ROC\_i = -\sin(3°) \cdot V\_i$$

For turboprops:

$$ROC\_i = -\sin(5°) \cdot V\_i$$

Also, compute F_i again using the following equations, update F_i, and proceed to main module 1 step 6.

$$\Delta t\_i = \frac{\Delta h\_i}{ROC\_i}$$

$$F\_i = \frac{m\_i \cdot g}{V\_i} \left(ROC\_i + \frac{V\_i}{g} \cdot \frac{\Delta V\_i}{\Delta t\_i} \right) + D\_i$$

Otherwise proceed to step 5.

5. Compute Δt [s] for each chord

If i = 1, $$\Delta t\_i = \frac{s\_i}{V\_i}$$

Otherwise, $$\Delta t\_i = \frac{\Delta h\_i}{ROC\_i}$$

SUB-MODULE D: DETERMINATION OF ENGINE FUEL FLOW AND FUEL BURN

1. Compute fuel flow [kg/s] for each chord

Compute percent power setting of engine thrust first.
   If F_i > 0 (positive thrust)

$$\%F_{oo}\_i = \frac{F\_i / N}{INMRO}$$

   If %F$_{oo}$_i > 1 or %F$_{oo}$_i < 0 below 3,000 ft at takeoff mode, %F$_{oo}$_i = 1.00
   If %F$_{oo}$_i > 1 or %F$_{oo}$_i < 0 at or above 3,000 ft at takeoff mode, %F$_{oo}$_i = 0.85
   If %F$_{oo}$_i > 1 or %F$_{oo}$_i < 0 at cruise mode, %F$_{oo}$_i = 0.30
   If %F$_{oo}$_i > 1 or %F$_{oo}$_i < 0 at approach mode, %F$_{oo}$_i = 0.07

If %F_i ≤ 0 (zero or negative thrust)

$$\%F_{oo}\_i = 0.07$$

If %F$_{oo}$_i ≤ 0.07,

   For jet and turboprop aircraft:

$$f\_i = \frac{C_{f3}}{60}\left(1 - \frac{3.2808 \times h\_i}{C_{f4}}\right)$$

For piston engines:

$$f\_i = \frac{C_{f3}}{60}$$

If %F$_{oo}$_i > 0.07 and h_i ≥ 7620 m (25,000 ft),

For jet aircraft:

$$f\_i = \frac{C_{f1}}{60000}\left(1 + \frac{1.9438 \times V\_i}{C_{f2}}\right) \cdot F\_i \cdot C_{fcr}$$

For turboprop aircraft:

$$f\_i = \frac{C_{f1}}{60000}\left(1 - \frac{1.9438 \times V\_i}{C_{f2}}\right) \cdot F\_i \cdot C_{fcr} \cdot \left(\frac{1.9438 \times V\_i}{1000}\right)$$

For piston engines:

$$f\_i = \frac{C_{f1} \cdot C_{fcr}}{60}$$

Otherwise,

For jet aircraft:

$$f\_i = \frac{C_{f1}}{60000}\left(1 + \frac{1.9438 \times V\_i}{C_{f2}}\right) \cdot F\_i$$

For turboprop aircraft:

$$f\_i = \frac{C_{f1}}{60000}\left(1 - \frac{1.9438 \times V\_i}{C_{f2}}\right) \cdot F\_i \cdot \left(\frac{1.9438 \times V\_i}{1000}\right)$$

For piston engines:

$$f\_i = \frac{C_{f1}}{60}$$

2. Compute fuel burn [kg] for each chord

$$FB\_i = f\_i \cdot \Delta t\_i$$

3. Debit aircraft mass [kg]

$$m\_i+1 = m\_i - FB\_i$$

If $m\_i+1 < 0.5 \times m_{ref}$,

$$m\_i+1 = 0.5 \times m_{ref}$$

SUB-MODULE E: DETERMINATION OF EMISSIONS

E-1. Mass balance to compute CO_2, H_2O, and SO_2 emissions [g] from fuel burn

$$CO_2\_i = 3155 \times FB\_i$$

$$H_2O\_i = 1237 \times FB\_i$$

$$SO_2\_i = 0.8 \times FB\_i$$

E-2. BM2 to compute HC, CO, and NO_x emissions

Note: Two types of curves need to be generated first in order to proceed with BM2. The steps 1 and 2 explain how to construct them.

1. Adjust and curve fit ICAO data

   a. Adjust W_f of ICAO 4 points to account for installation effects (engine air bleed).

   $$W_{ff\_TO} = 1.010 \times W_{f\_TO}, \quad W_{ff\_CO} = 1.013 \times W_{f\_CO}, \quad W_{ff\_AP} = 1.020 \times W_{f\_AP}, \text{ and } W_{ff\_ID} = 1.100 \times W_{f\_ID}$$

   b. Construct REIHC vs. W_{ff}, REICO vs. W_{ff}, and $REINO_x$ vs. W_{ff} curves using the adjusted data from step a. Use log-log scale for each of the pollutants with a linear fit for NOx and a bilinear fit for HC and CO.

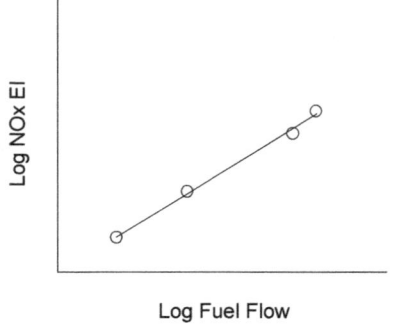
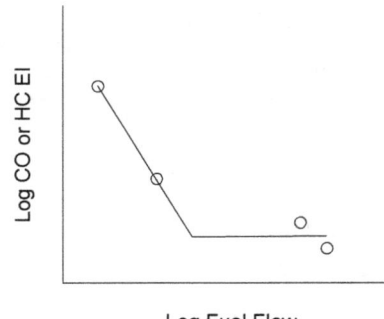

For the bilinear approach, a line is created between the two lower power setting points. The second line is horizontal and positioned in the middle of the two vertical positions of the higher power settings. Then the two lines are simply extended toward each other until they intersect.

2. Compute W_{ff_i}, $REIHC\_i$, $REICO\_i$, and $REINO_x\_i$ for each chord

   a. Compute temperature and pressure ratios.

   $$\boxed{\theta\_i = \frac{T\_i}{288.15}} \text{ and } \boxed{\delta\_i = \frac{P\_i}{101325}}$$

   b. Compute W_{ff}_i, which corrects W_f_i for altitude and Mach number.

   If $f\_i < W_{f\_ID}$,

   $$\boxed{f\_i = W_{f\_ID}} \text{ and } \boxed{W_{ff}\_i = \frac{f\_i/N}{\delta\_i} \theta\_i^{3.8} e^{0.2M\_i^2}}$$

   c. Determine $REIHC\_i$, $REICO\_i$, and $REINO_x\_i$ from $REIHC$ vs. W_{ff}, $REICO$ vs. W_{ff}, and $REINO_x$ vs. W_{ff} curves constructed in step 2 above. Use log scale for NO_x, HC and CO and bilinear fits for HC and CO.

3. Compute $EIHC\_i$, $EICO\_i$, and $EINO_x\_i$ for each chord [g/kg]

Note: $EIHC\_i$, $EICO\_i$, and $EINO_x\_i$ obtained here account for installation effects, altitude, Mach number, and humidity.

$$\boxed{EIHC\_i = REIHC\_i \frac{\theta\_i^{3.3}}{\delta\_i^{1.02}}}$$

$$\boxed{EICO\_i = REICO\_i \frac{\theta\_i^{3.3}}{\delta\_i^{1.02}}}$$

$$\boxed{EINO_x\_i = REINO_x\_i \cdot e^{H\_i} \cdot \left(\frac{\delta\_i^{1.02}}{\theta\_i^{3.3}}\right)^{1/2}}$$

$$\boxed{H\_i = -19.0\left(\frac{0.37318 \times P_v\_i}{P\_i - 0.6 P_v\_i} - 0.0063\right)}$$

$$\boxed{P_v\_i = 6895 \times 0.014504 \times 10^{\beta\_i}}$$

$$\boxed{\begin{aligned}\beta\_i = 7.90298\left(1 - \frac{373.16}{T\_i + 0.01}\right) + 3.00571 + 5.02808 \times log_{10}\left(\frac{373.16}{T\_i + 0.01}\right) + \\ 1.3816 \times 10^{-7}\left[1 - 10^{11.344\left(1 - \frac{T\_i + 0.01}{373.16}\right)}\right] + 8.1328 \times 10^{-3}\left[10^{3.49149\left(1 - \frac{373.16}{T\_i + 0.01}\right)} - 1\right]\end{aligned}}$$

4. Compute amount of emissions for each chord [g]

$$\boxed{HC\_i = EIHC\_i \cdot FB\_i}$$

$$CO\_i = EICO\_i \cdot FB\_i$$

$$NO_x\_i = EINO_x\_i \cdot FB\_i$$

SUB-MODULE F: DETERMINATION OF AIRLINE SPEED SCHEDULE

F-1. Determine transition altitude and initial speed for speed schedule

Use BADA APF files.

1. Compute δ_{trans}

$$\delta_{trans} = \frac{\left(1 + \frac{\gamma-1}{2}\left(\frac{0.51444 \times V_{cl\_2}}{a_o}\right)^2\right)^{\frac{\gamma}{\gamma-1}} - 1}{\left(1 + \frac{\gamma-1}{2}M_{cl}^2\right)^{\frac{\gamma}{\gamma-1}} - 1}$$

2. Compute h_{trans} [m] once for takeoff and once for landing

$$h_{trans} = \frac{T_o\_1 \times \left(1 - (\delta_{trans})^{0.19019}\right)}{0.0065}$$ for takeoff mode.

$$h_{trans} = \frac{T_o\_n \times \left(1 - (\delta_{trans})^{0.19019}\right)}{0.0065}$$ for landing mode (n denotes the last chord in landing).

3. Set initial mass [kg]

$$m\_1 = m_{ref} - FB\_taxiout$$

4. Set initial speed at head of first chord [m/s]

$$V\_1h = 0$$

F-2. Determine speed schedule for a given chord

Use BADA OPF files for V_{stall_TO} and V_{stall_LD} and APF files for all others.

1. Compute V_{CAS}_it and V_it [m/s]

Note: Speeds are computed at the head and tail of a chord and averaged in step 3 below to generate chord speed.

If joining chord or altitude $\geq h_{trans}$ (either takeoff or landing):

For takeoff, the tail speed of the joining chord is equal to the head speed (ground speed) of the first ETMS chord. Then calculate V_{CAS} back from the ground speed as following:

$$V_{CAS}\_it = \sqrt{\frac{2\gamma}{\gamma-1}\frac{101325}{1.225}\left\{1+\frac{P\_it}{101325}\left[\left(1+\frac{\gamma-1}{2\gamma}\frac{\rho\_it}{P\_it}(V\_it)^2\right)^{\frac{\gamma}{\gamma-1}}-1\right]^{\frac{\gamma-1}{\gamma}}-1\right\}}$$

For landing, the head speed of the joining chord is equal to the tail speed (ground speed) of the last ETMS chord. Then calculate V_{CAS} back from the ground speed as following:

$$V_{CAS}\_ih = \sqrt{\frac{2\gamma}{\gamma-1}\frac{101325}{1.225}\left\{1+\frac{P\_ih}{101325}\left[\left(1+\frac{\gamma-1}{2\gamma}\frac{\rho\_ih}{P\_ih}(V\_ih)^2\right)^{\frac{\gamma}{\gamma-1}}-1\right]^{\frac{\gamma-1}{\gamma}}-1\right\}}$$

Otherwise:

$$V_{CAS}\_i+1h = V_{CAS}\_it \quad \text{and} \quad V_{CAS}\_i = \frac{V_{CAS}\_it + V_{CAS}\_ih}{2}$$

$$V\_it = \sqrt{\frac{2\gamma}{\gamma-1}\frac{P\_it}{\rho\_it}\left\{1+\frac{101325}{P\_it}\left[\left(1+\frac{\gamma-1}{2\gamma}\frac{1.225}{101325}(V_{CAS}\_it)^2\right)^{\frac{\gamma}{\gamma-1}}-1\right]^{\frac{\gamma-1}{\gamma}}-1\right\}}$$

where:

   a. For takeoff,

   If $i = 1$, $V_{CAS}\_1t = 0.51444\left(1.2 V_{stall\_TO}\sqrt{\frac{m\_1}{m_{ref}}}+5\right)$ and $V_{CAS}\_2t = V_{CAS}\_1t$

   If $i = 3$, $V_{CAS}\_3t = V_{CAS}\_3h + 5.1444$

   If $i = 4$, $V_{CAS}\_4t = 0.51444\left(1.3 V_{stall\_TO}\sqrt{\frac{m\_4}{m_{ref}}}+60\right)$ and $V_{CAS}\_5t = V_{CAS}\_4t$

   If $i \geq 6$ and $h\_i < 3047.8$ m (10,000 ft), $V_{CAS}\_it = 128.61$

For all else, use the following speed schedule for takeoff.

h_it ≥ 3048 m (10,000 ft) to below h$_{trans}$: $\boxed{V_{CAS\_it} = 0.51444 \cdot V_{cl\_2}}$

h_it from and above h$_{trans}$: $\boxed{V\_it = M_{cl} \cdot a\_it}$ Note: This is not V$_{CAS}$; Rather, V$_{GRS\_it}$ is directly computed above h$_{trans}$.

Note: For a chord the attitude of which is greater than or equal to 1829m, pick one of the lo, av, or hi speed and Mach number values based on the mass of the aircraft at that chord (m_i). In order to do so, compute the differences between m_i and each of 1.2×m$_{min}$ (lo), m$_{ref}$ (av) and m$_{max}$ (hi). m_i then falls into the category that gives the least difference in mass.

b. For cruise, $\boxed{V\_it = M_{cr} \cdot a\_it}$

c. For landing,

h_it from and above h$_{trans}$: $\boxed{V\_it = M_{des} \cdot a\_it}$ Note: This is not V$_{CAS}$. Rather, V$_{GRS\_it}$ is directly computed above h$_{trans}$.

h_it from 3048 m (10,000 ft) to below h$_{trans}$: $\boxed{V_{CAS\_it} = 0.51444 \cdot V_{des\_2}}$

h_it from 1829 m (6,000 ft) to below 3048 m (10,000 ft): $\boxed{V_{CAS\_it} = 0.51444(min(V_{des\_1}, 250))}$

h_it from 914 m (3,000 ft) to below 1829 m (6,000 ft): $\boxed{V_{CAS\_it} = 0.51444(min(V_{des\_1}, 220))}$

h_it from 610 m (2,000 ft) to below 914 m (3,000 ft): $\boxed{V_{CAS\_it} = 0.51444\left(1.3V_{stall\_LD}\sqrt{\frac{m\_i}{m_{ref}}} + 50\right)}$

h_it from 457 m (1,500 ft) to below 610 m (2,000 ft): $\boxed{V_{CAS\_it} = 0.51444\left(1.3V_{stall\_LD}\sqrt{\frac{m\_i}{m_{ref}}} + 20\right)}$

h_it from 305 m (1,000 ft) to below 457 m (1,500 ft): $\boxed{V_{CAS\_it} = 0.51444\left(1.3V_{stall\_LD}\sqrt{\frac{m\_i}{m_{ref}}} + 10\right)}$

h_it from 0 to 305 m (1000 ft): $\boxed{V_{CAS\_it} = 0.51444\left(1.3V_{stall\_LD}\sqrt{\frac{m\_i}{m_{ref}}} + 5\right)}$

2. Compute V_ih [m/s]

Note: The speed we generate above is already in true airspeed. Therefore, there is no need for conversion.

$\boxed{V\_i+1h = V\_it}$

3. Compute ΔV_i and V_i [m/s]

$$\Delta V\_i = V\_it - V\_ih$$

$$V\_i(nw) = \frac{V\_it + V\_ih}{2}$$ where "nw" = no wind

If i = 1 or 2 (h_it <= 1000 ft), then
Per specifications in INM and SAE AIR 1845, make the following speed modification to include an 8 knot (4.11 m/s) headwind: $V\_i = V\_i(nw) - 4.11$

If i > 2, then
$V\_i = V\_i(nw)$

4. Compute M_i

$$M\_i = \frac{V\_i}{a\_i}$$

SUPPLEMENT 1: Chord Geometry

The geometry of a chord is defined by the following figure:

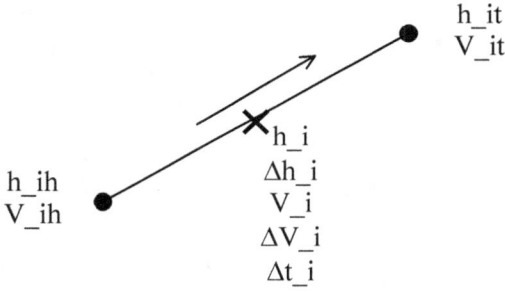

$$\Delta h\_i = h\_it - h\_ih$$
$$h\_i = (h\_ih + h\_it) / 2$$
$$\Delta V\_i = V\_it - V\_ih$$
$$V\_i = (V\_ih + V\_it) / 2$$
$$\Delta t\_i = actMiles / V\_i$$
$$ROC = \Delta h / \Delta t$$

The head and tail of a chord are referred to as nodes. As indicated by the "X," the rule of midpoint is used. All position, speed, time, ROC, thrust, fuel burn, emissions, and etc. associated with a chord are represented by the values at the midpoint of the chord.

One exception to this midpoint rule is aircraft mass, m_i. This is defined as the mass at the beginning of the ith chord and assumed to remain constant over the $\overline{i^{th}}$ chord segment and debited by the amount of fuel burned over i^{th} chord.

APPENDIX N: ICAO Engines Affected by Anomalous ICAO Data

The following engines are affected by anomalous ICAO data that do not adhere to the prescribed BFFM2 methodology. The issue numbers correspond to the issues discussed in Section 3.4.4.1. These issues will need to be addressed in an update of the BFFM2 writeup.

ICAO Engine Name	ICAO Engine UID	HC Plot Issues	CO Plot Issues	NOx Plot Issues	Fuel Flow Data Issues
D-30 (II series)	1AA001		Issue 1		
D-30KP-2	1AA002		Issue 1		
PS-90A	1AA005	Issue 1			
TFE731-2-2B	1AS001	Issue 1	Issue 1		
TFE731-2-2B	1AS001	Issue 1	Issue 1		
TFE731-3	1AS002	Issue 1	Issue 1		
TFE731-3	1AS002	Issue 1	Issue 1		
CF6-6D	1GE001		Issue 1		
CF6-6K	1GE003		Issue 1		
CF6-45A2	1GE005		Issue 1		
CF34-3A	1GE034		Issue 1, 6		
CF34-3A1	1GE035		Issue 1, 6		
NK-8-2U	1KK002			Issue 4	
NK-86A	1KK004	Issue 1			
JT3D-3B	1PW001		Issue 1		
JT3D-7 series	1PW002		Issue 1		
JT3D-7 series	1PW003	Issue 4	Issue 4	Issue 4	
JT8D-7 series	1PW004		Issue 1		
JT8D-9 series	1PW006		Issue 1		
JT8D-11	1PW008		Issue 1		
JT8D-15	1PW009		Issue 1		
JT8D-17	1PW012		Issue 1		
JT8D-17R	1PW016	Issue 1			
JT8D-209	1PW017	Issue 1	Issue 1		
JT8D-217 series	1PW018	Issue 1	Issue 1		
JT8D-219	1PW019	Issue 1			
JT9D-7	1PW020		Issue 1, 6		
JT9D-7A	1PW021		Issue 1, 6		
JT9D-7F	1PW023	Issue 1, 6			
JT9D-7J	1PW024	Issue 1, 6			
JT9D-7R4D, -7R4D1	1PW026	Issue 2			
JT9D-7R4E, -7R4E1	1PW027	Issue 2			
JT9D-20	1PW031		Issue 1, 6		
JT9D-20J	1PW032	Issue 1, 6			
JT15D-1 series	1PW035	Issue 1	Issue 1		
JT15D-4 series	1PW036	Issue 1	Issue 1		
JT15D-5, -5A, -5B	1PW037	Issue 1, 6	Issue 1, 6		
JT15D-5C	1PW038	Issue 1, 6	Issue 1, 6		
M45H-01	1RR001	Issue 4	Issue 1		
RB211-22B	1RR002		Issue 1		
RB211-22B	1RR003	Issue 1	Issue 1		
RB211-524B series	1RR004		Issue 1		
RB211-524C2	1RR006	Issue 1, 6	Issue 1		
RB211-524D4	1RR007	Issue 1, 6	Issue 1		
RB211-524G	1RR009	Issue 2			
SPEY Mk511	1RR015		Issue 1		
SPEY Mk555	1RR017		Issue 1, 6		
ALF 502R-3	1TL002		Issue 1		
ALF 502R-5	1TL003		Issue 1		
LF507-1F, -1H	1TL004		Issue 1		
D-36	1ZM001	Issue 4	Issue 4	Issue 4	Issue 4
CFM56-5B1/2	2CM016	Issue 3	Issue 3		
CFM56-5B2/2	2CM017	Issue 3	Issue 3		
CFM56-5B4/2	2CM018	Issue 3	Issue 3		
CF6-80C2A1	2GE036		Issue 1		
CF6-80C2A2	2GE037		Issue 1		
CF6-80C2A3	2GE038		Issue 1		
CF6-80C2A5	2GE039		Issue 1		
CF6-80C2A8	2GE040		Issue 1		
CF6-80C2B1	2GE041		Issue 1		
CF6-80C2B2	2GE042		Issue 1		
CF6-80C2B4	2GE043		Issue 1		
CF6-80C2B6	2GE044		Issue 1		
CF6-80C2B1F	2GE045		Issue 1		
CF6-80C2B2F	2GE046		Issue 1		
CF6-80C2B4F	2GE047		Issue 1		
CF6-80C2B6F	2GE048		Issue 1		
CF6-80C2D1F	2GE049		Issue 1		
CF6-80E1A1	2GE050		Issue 1		
CF6-80E1A2	2GE051		Issue 1		
GE90-76B	2GE052	Issue 1	Issue 1		
GE90-85B	2GE053	Issue 1	Issue 1		
CF6-80C2B7F	2GE055		Issue 1		

Trent 768	2RR022	Issue 1	
Trent 772	2RR023	Issue 1	
Trent 875	2RR024	Issue 6	
Trent 877	2RR025	Issue 6	
Trent 884	2RR026	Issue 6	
Trent 892	2RR027	Issue 2, 6	
BR700-710A1-10	3BR001	Issue 1, 6	
CFM56-5B1/2P	3CM020	Issue 3	Issue 1
CFM56-5B4/2P	3CM021	Issue 3	Issue 1
CFM56-5B6/2P	3CM022		Issue 1
CF6-80C2A5F	3GE056		Issue 1
CF6-80C2B5F	3GE057		Issue 1
CF6-80C2B8FA	3GE058		Issue 1
GE90-77B	3GE059	Issue 1	Issue 1
GE90-90B	3GE060	Issue 1	Issue 1
GE90-92B	3GE061	Issue 1	Issue 1
CF6-45A	3GE067		Issue 1
CF6-45A2	3GE068		Issue 1
CF6-50A	3GE069		Issue 1
CF6-50C	3GE070		Issue 1
CF6-50CA	3GE071		Issue 1
CF6-50C2R	3GE072		Issue 1
CF6-50C1	3GE073		Issue 1
CF6-50C2	3GE074		Issue 1
CF6-50E	3GE075		Issue 1
CF6-50E1	3GE076		Issue 1
CF6-50E2	3GE077		Issue 1
CF6-50C2B	3GE078		Issue 1
CF6-50E2B	3GE079		Issue 1
RB211-535E4	3RR028	Issue 1, 6	
Trent 768	3RR029	Issue 6	
Trent 772	3RR030	Issue 1, 6	
TAY 650	3RR031	Issue 2	
TAY 651	3RR032	Issue 2	
RB211-535E4B	3RR034	Issue 1, 6	
BR700-715A1-30	4BR002	Issue 1, 6	
BR700-715B1-30	4BR003	Issue 2, 6	
BR700-715C1-30	4BR004	Issue 6	
BR700-715A1-30	4BR005	Issue 2	
BR700-715B1-30	4BR006	Issue 2	
BR700-715C1-30	4BR007	Issue 2	
CFM56-5B2/2P	4CM037	Issue 3	Issue 1
CFM56-5B3/2P	4CM038	Issue 3	Issue 1
CFM56-7B22/2	4CM040	Issue 1	Issue 1
CFM56-7B24/2	4CM041	Issue 1	Issue 1
CFM56-7B26/2	4CM042	Issue 1	Issue 1
CFM56-7B27/2	4CM043	Issue 1	Issue 1
CF6-80E1A4	4GE081		Issue 1
JT8D-217	4PW068	Issue 7	Issue 1
JT8D-217A	4PW069	Issue 7	Issue 1
JT8D-217C	4PW070	Issue 7	Issue 1
JT8D-219	4PW071	Issue 7	Issue 1
RB211-524G-T	4RR036	Issue 2, 6	
RB211-524H-T	4RR037	Issue 2, 6	
CF34-8C1	5GE083	Issue 1	
CF34-3B	5GE084		Issue 2, 6
PW4X58	5PW074	Issue 6	Issue 1
PW4168	5PW075	Issue 6	Issue 1
PW4098	5PW076	Issue 7	
RB211-535E4	5RR038	Issue 1, 6	Issue 1, 6
RB211-535E4B	5RR039	Issue 1, 6	Issue 1, 6
Trent 895	5RR040	Issue 2, 6	
AE3007A1	6AL006	Issue 1, 6	Issue 1
AE3007A1	6AL007		Issue 1
AE3007A1/1	6AL009	Issue 1, 6	Issue 1, 6
AE3007A1/1	6AL010		Issue 1
AE3007A1/3	6AL012	Issue 1, 6	Issue 1, 6
AE3007A1/3	6AL013		Issue 1
AE3007A1P	6AL015	Issue 1, 6	Issue 1, 6
AE3007A1P	6AL016		Issue 1
AE3007A3	6AL018	Issue 1, 6	Issue 1, 6
AE3007A3	6AL019		Issue 1
AE3007A1E	6AL020		Issue 1
AE3007C	6AL022	Issue 1, 6	Issue 1, 6
AE3007C1	6AL024	Issue 1, 6	Issue 1, 6
CF34-8C5	6GE092	Issue 1	Issue 1
CF34-8E2	6GE093	Issue 1	Issue 1
CF34-8E5	6GE094	Issue 1	Issue 1
CF34-8E5A1	6GE095	Issue 1	Issue 1
CF34-8C1 Block 1	6GE096	Issue 1	Issue 1
Trent 556-61	6RR041	Issue 6	
TAY 611-8C	6RR042	Issue 2, 6	
CFM56-5C2/P	7CM045	Issue 6	
CFM56-5C3/P	7CM046	Issue 6	
CFM56-5C4/P	7CM047	Issue 6	
CFM56-5B8/P	7CM048	Issue 1	
CFM56-5B9/P	7CM049	Issue 1	
CFM56-5B9/2P	7CM050		Issue 1
PW306A	7PW077	Issue 6	
PW306B	7PW078	Issue 6	

PW308A	7PW079	Issue 1, 6	
PW4164	7PW081	Issue 6	Issue 1
PW4168A	7PW082	Issue 6	Issue 1

APPENDIX O: Forecasting Pseudo-Code

The following list of steps provides a pseudo-code style set of instructions to develop forecasted inventories of fuel burn and emissions.

CREATE BASELINE FLEET

1. Prepare a base year (e.g., 2000) fleet from the BACK database. This fleet is based on all aircraft in BACK that were listed as "active" (or in service) in that base year.

2. Read the BACK base year fleet records as aircraft objects and place into a collection:

   Object: Aircraft, with the following properties:

Property	Type	Status
Airline (Operator_Fullname)	String	Exists
AircraftType (Equipment_Type)	String	Exists
Age (Aircraft_Age)	Single	Exists
Seats (Total_Seats)	Integer	Exists
Country (Operator_Country)	String	Exists
SeatCategory	String	Create
OAGCountryCode	String	Exists
BACKOAGEquip	String	Exists
OAGEquip	String	Create
Count	Single	Create

   Collection: AircraftColBase

   Clean the base year fleet (e.g., remove any records with "NULL" or invalid data.

3. Assign a value of 1 to the Count property of each aircraft. This will later be modified based on aircraft retirements and replacements.

4. Assign each aircraft to a seat category (e.g., Aircraft.SeatCategory = 1).

   0. <=50
   1. 51-99
   2. 100-150
   3. 151-210
   4. 211-300
   5. 301-400
   6. 401-500
   7. 501-600
   8. >600

5. Prepare a list of the top 50 airlines with the 51st representing "the rest." Read and store the list of airlines which has already been developed by Paul Gerbi.

6. Categorize each of the aircraft objects in AircraftColBase by airline. This is accomplished by designating each of the aircraft to a specific airline or to the 51st airline. The airline category could be established using another property or rewritten over the Airline property.

7. Prepare a list of BACK to OAG equivalent aircraft types. Read in the records as objects (EquivAircraft.BACK and EquivAircraft.OAG) and place them into a collection (EquivAircraftCol).

8. Within each airline collection, categorize the aircraft objects by OAG aircraft types. This is accomplished by finding the equivalent OAG aircraft types from the objects in EquivAircraftCol and placing them in the OAGAircraft property of the Aircraft objects in the AircraftColBase collection.

9. Determine the number of each OAG aircraft type in each of the airline categories. These numbers represent the base year fleet and can be used for reference and debugging purposes.

CREATE FUTURE AIRCRAFT

10. Query the base year (e.g., 2000) BACK fleet database to obtain a list of pre-ordered aircraft (guaranteed orders) using the "On Order," "On Option," and "Pre-Service Built" labels. Store these aircraft as objects with the following properties:

Property	Type	Status
Airline (Operator_Fullname)	String	Exists
AircraftType (Equipment_Type)	String	Exists
Age (Aircraft_Age)	Single	Exists
Seats (Total_Seats)	Integer	Exists
Country (Operator_Country)	String	Exists
SeatCategory	String	Create
OAGAircraft	String	Create
PhaseOutIndicator	String	Create
PhaseOutFraction	Single	Create

These aircraft will be used as "state-of-the-art" aircraft for replacement purposes.

11. Assign equivalent OAG equipment types to each of the aircraft.

12. Assign seat categories to each aircraft as follows:

Seat Category	Seat Range
0	<=50
1	51-99
2	100-150
3	151-210
4	211-300
5	301-400
6	401-500
7	501-600
8	>600

13. Develop fractional distributions of these aircraft in each seat category. However, assign the most popular aircraft for seat categories 0 and 1. Review each of these future aircraft to make sure they "make sense" (are plausible) as "state-of-the-art" aircraft. Remove any that are questionable based on expert judgment and renormalize the distributions.

CREATION OF FORECASTED FLEETS (ONLY GROWTH)

14. Copy all aircraft in AircraftColBase and put them into a new collection, GrowCol, and use this as the basis to create future fleets.

15. Age each aircraft by one year. This provides a "raw" (with no retirements and substitutions) base year+1 fleet with correct aircraft ages.

16. Prepare the FESG growth projections for the base year and the forecast year. Then read the records as objects (FESG.SeatCategory, FESG.Base, and FESG.Forecast) and place into a collection (FESGCol).

AcNumber	2000	2001	2002	---	2020
51 - 99	1646	1696	1758	---	2419
100 – 150	5752	5902	6067	---	7199
151 – 210	1686	1747	1820	---	4254
211 – 300	1117	1163	1217	---	2572
301 – 400	619	646	675	---	1141
401 – 500	83	92	105	---	447
501 – 600	24	26	29	---	375
601 – 650	0	0	0	---	471

17. Develop incremental yearly growth factors for the FESG projections and place them into a property (e.g., FESG.GrowthFactor).

18. Determine the number of new aircraft (due to FESG growth projections) by multiplying the count property of each aircraft with the corresponding growth factor based on seat category.

    IF Aircraft.SeatCategory = FESG.SeatCategory THEN
       Apply appropriate growth factor
    END IF

    This provides the number of new aircraft objects to be added to GrowCol. Assume aircraft in SeatCategory = 0 is the same as those in SeatCategory = 1 since FESG projections for seat categories below 51 are not provided.

19. Read and store the list of countries with the noise chapter 2 phase out rule designations (i.e., Y or N). Using this information and each aircraft's Country property, assign a Y or N to each aircraft's PhaseOutIndicator property.

20. Assign a "phase out" fraction to each of the aircraft not affiliated with a country that has adopted the noise phase out rule (i.e., aircraft.PhaseOutIndicator=N):

2002	0.63
2006	0.67
2013	0.88
2020	0.96

An interpolation scheme will need to be developed for the in-between years.

21. In order to determine the new aircraft type, the noise phase out data from Steps 19 and 20 must be used as follows:

    - If the airline has adopted the noise phase out rule (i.e., aircraft.PhaseOutIndicator = Y), then all new aircraft (i.e., all grown fractions) will be based on on-order aircraft types.

    - If the airline has NOT adopted the noise phase out rule (i.e., aircraft.PhaseOutIndicator = N), then use the appropriate phase out fraction from Step 20 to determine the fraction of the new aircraft that will be based on on-order aircraft and the fraction of the new aircraft that will be based on the existing aircraft:

    Fraction of new aircraft based on on-order aircraft = Fn * Pf
    Remaining fraction of new aircraft based on existing aircraft = 1 - (Fr * Pf)
    where Fn = count property of new aircraft
         Pf = Phase out fraction based on country property

    The newly created aircraft will have the same properties as those they were grown from. However, assign an age of 0.5 to these new aircraft since the delivery month is unknown. Assign a value of 1 for each whole aircraft and a fraction for each non-whole aircraft to the Count property.

22. Repeat steps 15-19 consecutively by year until the forecast year has been reached. The aircraft objects in GrowCol only experience growth effects (no retirements or replacements).

APPLICATION OF RETIREMENT PARAMETERS

23. Create a copy of GrowCol and call it GrowRetCol. This new collection of aircraft will be used to retire and replace aircraft.

24. For each aircraft, apply the FESG survivor curves to determine the fraction of aircraft that survived.

    $S = \text{constant} + ax + bx^2 + cx^3 + dx^4 + ex^5 + fx^6$
       = survival factor (fraction of aircraft that survived)
    x = age of aircraft

	Curve 1	Curve 2	Curve 3	Curve 4
Constant	0.7912	0.875867	0.277046	0.782491
a	0.0975	0.039574	0.136525	0.080313
b	-0.016835	-0.00352285	-0.0076598	-0.00931738
c	0.0013517	0.0000478103	0.000103682	
d	-0.000053636			
e	-0.00000097731			
f	-6.581E-09			

| Range | 7 to 47 years | 7 to 36 years | 12 to 36 years | 5 to 14 years |

Curve 1: All else. Larry Gray says: "Newer generation aircraft plus the old 2-man flight crews such as 737-200 and DC9s."
Curve 2: Larry Gray says: "1st generation wide body aircraft (A300B4, L1011, DC10, 747-100/200/300)."
Curve 3: Larry Gray says: "B727s and B707s."
Curve 4: Larry Gray says: "MD-11."

Instead of using these equations, a pre-generated table of survivor factors (for each year) could be used. Using the survivor factors, the retirement fraction is calculated as follows:

Fr = 1 – S = Fraction of aircraft to retire.

25. In order to determine the replacement aircraft type, the noise phase out data from Steps 19 and 20 must be used as follows:

- If the airline has adopted the noise phase out rule (i.e., aircraft.PhaseOutIndicator = Y), then replace all retired aircraft with on-order aircraft.

- If the airline has NOT adopted the noise phase out rule (i.e., aircraft.PhaseOutIndicator = N), then use the appropriate phase out fraction from Step 20 to determine the fraction of the retired aircraft that will be based on on-order aircraft and the fraction of the retired aircraft that will be based on the existing aircraft:

    Retirement fraction to be replaced with on-order aircraft = Fr * Pf
    Remaining fraction to be replaced with existing aircraft type = 1 - (Fr * Pf)
    where Fr = Fraction of an aircraft to retire as previously defined
        Pf = Phase out fraction based on country property

The replacement aircraft objects should have the same properties as those they replaced. Assign an age of 0.5 to these new aircraft since the delivery month is unknown. Assign a value of 1 for each whole aircraft and a fraction for each non-whole aircraft to the count property.

DEVELOPMENT OF CONVERSION FACTORS

26. Determine the number of aircraft in each airline and OAG aircraft category for the forecast year by using the Count property.

27. Determine conversion factors as follows:

    CF = Conversion Factor = nGR / nG
    where nG = number of aircraft in each category due to growth only (from GrowCol)
        nGR = number of aircraft in each category due to growth and retirement (from GrowRetCol)

The conversion factors reflect the effects of just the retirements and replacements.

DEVELOPMENT OF FUTURE SCHEDULES

28. Prepare the OAG flight schedules, the FESG projections, and Terminal Area Forecasts (TAF) and other supporting data.

29. Run the "Fesg2.sas" script to develop growth rates based on the forecasted projection data in Step 28.

30. Run the "Schedule13.sas" script to obtain a week's worth of grown OAG flight schedules. This script uses the TAF with a Fratar algorithm to distribute future projections of aircraft operations to flights in the US, and uses the FESG-based growth factors to grow flights in other regions. The resulting future schedules will reflect the growth in flights by replicating existing flights and adding them to the schedules. Ideally, these additional flights should have non-whole number multipliers associated with them to reflect the growth rates. But since these schedules are used in WWLMINET, only individually listed whole number flights (i.e., "tickets") could be used. Therefore, a heuristic algorithm was developed to round the results to whole numbers.

31. Feed the schedules (and the derived airport demand information) into WWLMINET to obtain airport delays for each of the 257 worldwide airports and flight cancellation information.

32. Using the flight cancellation information from WWLMINET, remove the canceled flights from the future schedules and aggregate the individual listings of grown flights with the original (base) flight. The aggregation of the flights should be reflected by an additional flight multiplying (FM) factor.

APPLICATION OF CONVERSION FACTORS

33. Multiply the conversion factors from Step 27 with the flight multiplying factors from Step 32 based on airline and aircraft type categories used in developing the conversion factors. This will generate a new flight multiplier (NFM) for each flight listed in the aggregated schedule from Step 32.

    NFM = New Flight Multiplier = CF * FM

    The new flight multiplier takes into account all of the effects of growth, retirement, and replacement.

    Although a disconnect exists between the growth of the fleet and the growth of the schedules, the application of the conversion factors to the schedules is justified because the growth of the schedules (from the FESG flight projections) took into account aircraft seat size and range. Therefore, the relative strengths of the conversion factors by aircraft type will generally correspond to the relative strengths of the flight multipliers by aircraft type. That is, the aircraft with the higher conversion factors will be reflected by a similar prominence of flights (as reflected by the flight multipliers) for the same or similar aircraft types.

34. If an aircraft type for a certain carrier from the fleet data has no match in the schedule, additional fights are created for this aircraft by:

    - Prepare a table of aircraft counts for each airline including the 51st generic airline.

    - Determine the ratio of counts for the aircraft in question with the most popular (highest count) aircraft in the same airline and range categories. This ratio should be less than 1 unless the aircraft in question is the most popular, in which case the ratio will be based on a comparison with the next most popular. If an airline match cannot be made, use the 51st category.

- Create additional flights in the schedule by copying flights with the most popular aircraft (or the next most popular) and multiplying the ratio with the new flight multiplier (NFM).

This may be due the fact that some on-order aircraft currently has no flights listed in the base year schedule. The additional flights need to be created in order to preserve the aircraft counts created during the aging and retiring process.